STATISTICS IN SCIENCE

BOSTON STUDIES IN THE PHILOSOPHY OF SCIENCE

VOLUME 122

STATISTICS IN SCIENCE

The Foundations of Statistical Methods in Biology, Physics and Economics

Edited by

ROGER COOKE

Department of Mathematics, Delft University of Technology, The Netherlands

and

DOMENICO COSTANTINI

Institute of Statistics, University of Genoa, Italy

KLUWER ACADEMIC PUBLISHERS
DORDRECHT / BOSTON / LONDON

Library of Congress Cataloging-in-Publication Data

Statistics in science : the foundations of statistical methods in
 biology, physics, and economics / edited by Roger Cooke, Domenico
 Costantini.
       p.   cm. -- (Boston studies in the philosophy of science)
   Proceedings of the International Conference on Statistics, held in
 Luino, Italy; organized by the Società italiana di logica e
 filosofia della scienza and the Istituto Ludovico Geymonat.

   1. Research--Statistical methods--Congresses.   2. Science-
-Statistical methods--Congresses.   I. Cooke, Roger.
 II. Constantini, Domenico.   III. International Conference on
 Statistics (Luino, Italy)   IV. Società italiana di logica e
 filosofia della scienza.   V. Istituto Ludovico Geymonat.
 VI. Series.
 Q180.55.S7S73   1990
 001.4'22--dc20   90-36340

ISBN-13: 978-94-010-6765-2   e-ISBN-13: 978-94-009-0619-8
DOI: 10.1007/978-94-009-0619-8

Published by Kluwer Academic Publishers,
P.O. Box 17, 3300 AA Dordrecht, The Netherlands.

Kluwer Academic Publishers incorporates the publishing programmes of
D. Reidel, Martinus Nijhoff, Dr W. Junk and MTP Press.

Sold and distributed in the U.S.A. and Canada
by Kluwer Academic Publishers,
101 Philip Drive, Norwell, MA 02061, U.S.A.

In all other countries, sold and distributed
by Kluwer Academic Publishers Group,
P.O. Box 322, 3300 AH Dordrecht, The Netherlands.

Printed on acid-free paper

TABLE OF CONTENTS

INTRODUCTION

An *inference* may be defined as a passage of thought according to some method. In the theory of knowledge it is customary to distinguish *deductive* and *non-deductive* inferences. Deductive inferences are *truth preserving*, that is, the truth of the premises is preserved in the conclusion. As a result, the conclusion of a deductive inference is already 'contained' in the premises, although we may not *know* this fact until the inference is performed. Standard examples of deductive inferences are taken from logic and mathematics. Non-deductive inferences need not preserve truth, that is, 'thought may pass' from true premises to false conclusions. Such inferences can be *expansive*, or, *ampliative* in the sense that the performances of such inferences actually *increases* our putative knowledge. *Standard* non-deductive inferences do not really exist, but one may think of elementary inductive inferences in which conclusions regarding the future are drawn from knowledge of the past.

Since the body of scientific knowledge is increasing, it is obvious that the method of science must allow non-deductive as well as deductive inferences. Indeed, the explosive growth of science in recent times points to a prominent role for the former. Philosophers of science have long tried to isolate and study the non-deductive inferences in science. The inevitability of such inferences one the one hand, juxtaposed with the poverty of all efforts to identify them, constitutes one of the major cognitive embarrassments of our time.

The reasons for compling a book on a subject as expansive as 'Statistics in Science' can be traced to the conviction that all non-deductive inferences in science are ultimately statistical inferences. In other words, a non-deductive inference in science ultimately reduces to drawing conclusions about the degree to which hypotheses are supported by data. Moreover, this support must be probabilistic in character and cannot be adequately formalized without making use of probability theory.

As is well known, the first modern use of probability was related to problems of gambling and insurance involved with predicting future

events. Such predictions are made by stating a probability of occurrence for the events in question, on the basis of certain probabilistic
hypotheses (e.g. independence in the case of gambling, known relative
frequencies in the case of insurance).

T. Bayes enlarged the statistical methodology by considering the
probability of a probability distribution. P. S. Laplace and C. F. Gauss
introduced this method in science making use of prior probabilities and
evidencs. J. C. Maxwell, L. Boltzmann, R. Galton and K. Pearson, being
unable to specify any prior distribution for the problems they were
treating, gave up the method envisaged by Bayes and succeeded in
adapting the Galilei hypothetical-deductive methods to cases where the
hypotheses are statistical in character. The kinetic theory of gas and the
'objectivist' theory of statistical testing are the splendid results of their
efforts.

The two dominant streams in statistical methodology, Bayesian and
objectivist, cover most of what can be called 'data analysis', i.e. trying to
determine what conclusions to draw from data. However, the province
of non-deductive inference has become much wider than data analysis.
For example, the multivariate techniques originally introduced for data
reduction are now used as an exploratory tool in genetics, economics
and social science. Subjective probability was originally introduced to
draw conclusions about statistical hypotheses from data, but is becoming
very widely used in the form of *expert opinion* in quantitative risk
analysis, policy analysis and mathematical decision support. Probabilistic symmetry originally introduced by B. de Finetti to explain the
prominent role of relative frequencies in probabilistic reasoning, has
become an important tool in understanding the behavior of elementary
particles in quantum physics. In all these cases, 'probabilistic reasoning'
has wandered off the reservation of mainstream statistics and has
become an integral part of diverse scientific disciplines.

Perhaps the most dramatic example of this is the role of probability
itself in quantum mechanics. In appropriating probabilistic notions,
quantum theory wrought radical changes both in the formalism and in
the interpretation of probability. The extent of these changes is a
subject of a long and very rich discussion which is still in progress.

The International Conference on Statistics in Science was not
organized in an attempt to bring probabilistic thinking back on the
reservation of standard statistical reasoning. On the contrary it is
believed that the proliferation of quantitative probabilistic methods in

diverse scientific disciplines is salutory and inevitable. However, proliferation makes it difficult to keep track of developments in diverse disciplines, and probalistic thinking runs the risk of becoming fragmented. Those who have experienced the power of analogy in probabilistic reasoning know how high the costs of such fragementation might be. Researchers in scientific disciplines who employ probalistic methods in methodologically innovative must communicate with their brethern in other discipolines.

The conference in Luino was intended to be a first step in this direction. We cannot claim that the contributors from econometrics, game theory, risk analysis, population genetics, biology, and quantum physics embraced each other as long lost family. However, there was a clear sense of common purpose. Everyone *wanted* to know what the others were doing, and everyone came away knowing more than he/she knew before. Moreover, there were certain common themes which emerged repeatedly. One such theme was the notion of symmetry, or exchangeability, which seems to turn up in unlikely places. Another theme was the recurrent need to codify and justify procedures. Indeed, the strongest force driving the fragmentation of our discipline is the pressure to get results within a pre-defined time frame. One does not have the time to reflect on interesting methodological questions whose bearing on the application at hand is secondary. Hence, one takes recourse to ad hoc procedures. These adhockeries accumulate, get canonized by default, and subsequently pose formidable barriers to communication.

The organizers of the conference in Luino hope that these proceedings will give the reader a flavor of the atmosphere of the conference, and will help to establish a unified vision of probabilistic reasoning in science. The Luino Conference was organized by the Societá Italiana di Logica e Filosofia delle Scienze and the Istituto Ludovico Geymonat. It was supported by Acciaiere e Ferriere Vicentine Beltrame S.p.A, Banca Popolare di Luino e di Varese, Comunita Montana Valli del Luinese and Comune di Luino. Moreover the Comune di Luino gave hospitality to the Conference in the Palazzo Verbania thus ensuring a marvelous venue.

DOMENICO COSTANTINI
ROGER COOKE

LORENZ KRÜGER

METHOD, THEORY, AND STATISTICS:
THE LESSON OF PHYSICS

INTRODUCTION

The subtitle of my paper — The Lesson of Physics — is an overstatement in several ways: *The* lesson is, of course, no more than the lesson *I* have extracted from the history of physics for myself, and that I would like others to believe as well. Hence, it is only one lesson among several possible lessons, i.e. a view suggested for discussion. Moreover the singular 'the lesson' requires a historical survey of about 200 years of physics and a systematic analysis of its present results. Needless to say, then, that I rely heavily on the work of others, among them several participants of this conference.

When we talk about knowledge, especially scientific knowledge, we imply that the things of which we claim knowledge are as we say they are; and if we say something well-defined, we appear to assume that they are in a certain well-determined way. If, then, something sometimes happens in this way and sometimes in another, as chance would have it, we appear to be deprived of knowledge about that thing. This traditional opposition of chance and knowledge was generally accepted when modern science emerged. There are even pictures confronting two allegoric figures *Fortuna* and *Sapientia*. A frontispiece of a 16th century edition of Petrarch, for example, shows Fortuna on the left, blindfold, holding her wheel and seated on a round, hence unstable seat. Facing her, Sapientia is firmly established on a square and stable throne; not only can she see, she can even become certain of herself, since she is equipped with a mirror that reflects her face.

It is one of the surprising achievements of modern science, if not its most important achievement, that it has brought even the accidental under its control. In the terms of the allegory, it has merged Fortuna and Sapientia into one figure: Scientia Statistica. How has that feat, undreamed of in Antiquity and the Middle Ages, been made possible? One necessary condition has been the discovery of the concept of probability in the second half of the seventeenth century.

Although the concept of probability and modern physics emerged in the same century, they were by no means natural allies. On the con-

R. Cooke and D. Costantini (eds), Statistics in Science. The Foundations of Statistical Methods in Biology, Physics and Economics, 1—13.

trary, physics soon became the most powerful instantiation of the Sapentia of our allegoric picture. Modern physics, if any science, acquired its fame as the science of the strict law-governed order of nature. Under the Newtonian program of classical mechanics the material world is viewed as matter in motion under the impact of forces that completely determine its course.

How then did physics turn from a bullwark of determinism into a bridgehead of statistical science? This drama, as fits a classical play, has five acts. For brevity let me label them as follows:

1. the statistics of observations,
2. the statistical description of complex systems,
3. the statistical theory of irreversibility,
4. the statistics of elementary processes,
5. quantum statistics.

(A more detailed account of 1. through 4. will be found in: Gigerenzer *et al.*, 1989, Chapter 5.)

The drama turned into a tragedy for the metaphysical religion of determinism. In the first act it still looks as if statistics were marginal and a mere methodological tool; but in the end statistical structures are found at the heart of matter. Of course, the drama is as well known as, say, *King Lear*; but it is not viewed as the defeat of determinism by all beholders. I shall, therefore, review some of its key scenes and try to expose those episodes during which, as it were, the statistical plot thickens.

1. THE STATISTICS OF OBSERVATION

Repeated measurements of what is supposed to be one and the same magnitude as a rule do not agree with each other. How are they to be combined? The question arose early on in the history of modern science, i.e. around 1600 (Eisenhart, 1971). But it took a while — roughly a century and a half — to recognize that it should be solved by investigating the form of the statistical distribution of measured values. The actual problem facing, say, the astronomer was still more complicated, because often the data had to be related to an entire series of interconnected magnitudes, e.g. to the successive positions of a planet along its path. Around 1800 Legendre and Gauss independently found the solution that has since been generally accepted: the method of least squares. It was Gauss (1809) who showed that this method can be justified on the assumptions that ever larger discrepancies between the true and the observed values of a magnitude become ever more improbable, and that the arithmetical mean of the observed values is the most probable estimate of the true value. This argument marks the beginning of scientific statistics in physics, i.e. a statistics guided by theoretical understanding.

Laplace, in 1810, clarified the accidental statistical character of the error curve further by showing that a sum of many independent errors will always be normally distributed (cp. Sheynin, 1977). His result was applied by Hagen (1837) and Bessel (1838), who explained the normal or Gaussian distribution by arguing that each single observable error is itself a superposition of many independent invisible elementary errors — an idea first suggested by Thomas Young in 1819.

Thus it happened that statistics took a firm hold in physics. But it did not seem to threaten the deterministic mechanical ideal, since it was strictly confined to the level of *method*. Provided that the observational errors were not systematic, as they are for instance in the so-called 'personal equation' for the individual observer, they were not taken seriously as natural phenomena. The purpose of statistics was to ascertain the one true and precise value of each magnitude, a value whose existence appeared to be guaranteed by the available theories. Physicists had good reasons to believe that their theories would always be strong enough to tell them what structure they were to expect in their objects. Hence in physics, as opposed to the social sciences, statistics remained auxiliary for quite a while. This changed only when error as the subject matter of the statistical distributions was replaced by something more physical and more substantial. That occurred in 1860 when Maxwell announced that the velocities of molecules in a gas were also distributed according to the law of error. His first paper on the kinetic theory of matter opens the second act of our drama.

2. THE STATISTICAL DESCRIPTION OF COMPLEX SYSTEMS

The basic plot of this second act is very simple. In a nutshell, it runs thus: The mechanical program in combination with the atomistic conception of matter results in a statistical theory. Simple as this plot may be, it is not obvious how one ought to assess its epistemological, or possibly even ontological, impact. It is build into the classical mechanical approach that it serves the deterministic ideal only to the extent that a full and fully determined description of all relevant details of a mechanical system can be attained for a given time. Now, it is obvious that this requirement cannot be fulfilled any more as soon as the atomistic constitution of matter becomes relevant for the explanation of observable phenomena, as it did, for instance, in the case of thermal and chemical phenomena in the second half of the 19th century.

How should we describe the epistemological status of the statistical theory of matter as developed mainly by Maxwell and Boltzmann? Not only are we *incapable* to follow the dynamical history of a single molecule, let alone of all individual molecules, we are also *uninterested* to do so. The atomistic and molecular constitution of matter is assumed in order to explain a range of qualitatively different macroscopic phenomena. Now, the very success of this explanatory program requires that we correlate an enormously large number of different mechanical microstates with each single distinguishable macrostate. In other words, it belongs to the nature of this explanatory theoretical attempt that large sets of alternative microstates must somehow be dealt with simultaneously. And here, these states are alternative precisely in the sense that they do *not* belong to a common dynamical history; on the contrary, only one of them can be actual. In this sense, then, statistical theory is entirely different from dynamical theory. The connection between these two theoretical approaches consists only in the fact that the statistics is a statistics of the dynamical magnitudes as defined in the framework of traditional mechanics.

The additional principles of the statistical theory, however, do not follow from mechanics; nor do they follow from observation. They are independent assumptions whose validity is tested by examining the experimental data that are derivable from them. As a matter of historical record, this is the correct description. At least, it is largely correct up to the present. It is true that since Maxwell's last paper on the kinetic theory of matter (1878) there was the idea and the program of deriving those statistical assumptions from the dynamical history of the single real system, i.e. there was the ergodic hypothesis and later also the impressive and rapidly expanding field of ergodic theory (see e.g. von Plato, 1987). But applications to realistic physical cases are at best in its early beginnings. It seems therefore, appropriate to view the ergodic program as one of a reconciliation between the traditional mechanical picture and the more recent statistical principles, principles however that had long proved their mettle on independent grounds.

What this means to the epistemologist is that we are entitled to consider the following question to be open: Does mechanics provide the adequate picture of macroscopic matter? Or does an adequate account perhaps require additional concepts which only make sense in the context of a more abstract theory, a theory that deliberately ignores a complete description of elementary processes in space and time? Temperature, for instance, is a good candidate for such a concept. And after we have been taught by quantum theory that the microworld resists visualization in classical terms anyhow, the suggestion of a theory violating some of the traditional pictorial requirements need no longer be looked upon as still somehow defective. In other words: it is not clear any more that the basic constitution of matter can be spelled out in terms of a small set of manifest properties, preferably all primary properties in the common philosophical sense of that term. Moreover, at the end of the 19th century, mechanics could not be taken any more to support philosophical determinism, as it clearly could around 1800.

Nevertheless, one might well say, classical mechanics had not (yet) been violated; it had just been complemented by additional assumptions. Indeed, the hero of the drama, mechanical determinism, was still well and alive; his rule remained unbroken in most parts of the scientific empire. The thrid act, however, will show how dangereously this dominion was undermined already then.

3. THE STATISTICAL EXPLANATION OF IRREVERSIBILITY

There are a number of attempts at explaining the pervasive irreversibility of our world: some cosmological, others related to the decay of elementary particles. But the overwhelming majority of irreversible processes of our daily experience, like mixing milk and coffee or preparing a well-tempered bath from hot and cold water, proceed according to the second law of thermodynamics. Although this law had been a major explanandum of the kinetic theory of matter from its beginnings around 1860, it remained its stumbling stone. Maxwell's statistical physics dealt only with equilibrium states; ergodic theory was and is similarly restricted. Ludwig Boltzmann struggled with the mechanical explanation of irreversibility during his whole life. In the end he scored at best a partial success; but — though this is still a matter of ongoing research and debate — it may be claimed that he pointed into the right direction.

In 1872 Boltzmann, on the basis of an ingenious treatment of molecular collisions, succeeded in defining a certain function of the molecular velocities, which he called H and which always decreases with time until it reaches a minimum. Its negative, therefore, could be identified with the macroscopically defined thermodynamic function of entropy. At first Boltzmann believed to have given 'a strict proof' of the entropy law on purely mechanical grounds (1872, p. 345). But well-known objections by his teacher Loschmidt and the mathematician Zermelo showed that such a proof must be faulty. For the laws of mechanics are strictly symmetric with respect to time, so that irreversible phenomena could not possibly be derived from them. In our context, the details of this story cannot and need not be told again. (An excellent account is found in Brush, 1976).

For my present argument it is enough to remember one pivotal feature of Boltzmann's defense (1877): He showed that states with low H-value, i.e. high entropy, can be realized by overwhelmingly many microstates, whereas states with high H-value, or low entropy, are realized by comparatively few microstates. He then interpreted the respective numbers of possible microstates as the probability of the corresponding macrostates. Finally, he argued that a system will always move from less probable to more probable states. This argument is ingenious and persuasive. The trouble is only that it bypasses all dynamical considerations. For, as long as no dynamical connection between successive states is taken into account, there is no less reason to argue that the system *has been* in a more probable state in the past than that it *will be* in a more probable state in the further. Hence, the source of irreversibility must lie elsewhere. Eventually, Boltzmann turned to cosmological speculations.

In view of this result of his life-long effort, we are led to ask how he could ever have obtained his irreversible equation for H, now called the 'Boltzmann equation'. The answer is simple: In his equation the ordinary mechanical magnitudes, positions and velocities, have been replaced by statistical distributions of such magnitudes; only the latter enter into dynamical relationships. In other words, in order to obtain irreversible processes, one needs a dynamical equation that is itself asymmetric with respect to time, and this type of equation required statistical magnitudes rather than individual mechanical magnitudes. Ilya Prigogine and his followers have taken this lesson seriously. Prigogine argues that time-irreversible phenomena demand a completely new conception of dynamics (1984, VII.1, X.1) based on distribution functions as basic magnitudes rather than on positions and momenta (1962, p. 6). It is not for the philosopher to say whether or not Prigogine's program will finally succeed. But this program is certainly a good reason to direct our philosophical curiosity towards the question of what the deeper connection is between the statistical characterization of moving matter on the one hand and the irreversibility of the motion on the other.

A question like this one may sound less strange when we see it against the background of the statistical constitution of matter as discovered in quantum physics, to which I now turn.

4. THE STATISTICS OF ELEMENTARY PROCESSES

Until about 1900, or so one might argue, the use of statistics in physics was motivated either by methodological convenience, or by certain goals of theoretical explanation, or finally by a (possibly temporary) incapability of reconciling mechanics with irreversibility without the use of statistical theory. Thus a purely epistemic interpretation of statistics became very common among physicists. It is indeed supported by the fact that all statistical phenomena discussed so far are mass phenomena, i.e. they appear only in systems consisting of many different parts. It is therefore possible to assume that each part taken in isolation has nothing to do with statistics.

This situation changed fundamentally shortly after 1900 when it became clear that radioactive decay is most naturally explained in terms of disintegration probabilities. Soon after the discovery of radioactivity Pierre and Eve Curie, Rutherford and others showed that the decay rate is not affected by any circumstance outside the instable atoms, especially that it does not depend on the mutual interaction of those atoms. The uniform statistical behaviour of the molecules of a gas had been assumed to result from the innumerable collisions that rapidly and effectively redistribute the mechanical properties of the individual particles. But this explanation was barred in the case of radio-atoms, so that, in 1903, Rutherford and Soddy announced the idea of a constant decay probability per unit of time characterizing *each individual atom*. Soon thereafter the consequences of this assumption were worked out: there must be chance fluctuations of actual decay rates and of the time intervals between two successive distintegrations. All this was very nicely confirmed by experiment.

Thus the combination of the following two circumstances created an entirely new rôle for statistical ideas in physics:

1. the immediate visibility of statistical fluctuations, and
2. the impossibility of explaining them as mass phenomena.

The disintegration probabilities inherent in individual atoms were a new kind of source of statistical phenomena.

Of course, radioactive decay by itself would not have been sufficient to secure, indeed even to recognize, this new source. This recognition emerged only gradually as quantum physics grew, and was only completed with the event of quantum mechanics around 1926. In 1928 also radioactive decay was explicitly integrated into the new theory (Gamow, 1928; Gurney and Condon, 1929). In this theory the discovery that had first been made with respect to radioactivity was generalized to all phenomena on the atomic level: Statistical appearances emerge from the constitution of individual systems or from the nature of individual processes. That means: Although statistics as a *phenomenon* requires by its very concept a sufficiently large number of observable cases, its *cause* is not longer sought in the overall structure of the large assembly of those cases, but in the nature of each and every individual part of the assembly alike. Quantum mechanics has taught us the idea that it may belong to the internal constitution of an elementary physical system to display a statistical pattern of actions or reactions instead of a uniquely determined behaviour under given circumstances.

Many philosophers, perhaps also some scientists, will protest against the realistic interpretation of statistical patterns implicit in my description. But I hope to have at least indicated, if not sufficiently explicated, why this interpretation can hardly be avoided, once two points have been accepted:

a) that quantum mechanics is correct, at least in the relevant respects; and
b) that the elementary systems it deals with, e.g. atoms or nuclei, are as real as stones and stars, though they certainly are different kinds of real things with some rather strange properties.

One of those strange properties belongs in my story, since it has to do with statistics. Its discovery is the fifth and last of act of our drama.

5. QUANTUM STATISTICS

So far we have only considered a completely general feature of quantum mechanics: it unexceptionally characterizes its objects by probabilities, hence by patterns of statistical behaviour. Yet, statistics enters quantum theory in a much more specific and surprising way whenever a system contains two or more particles of the same kind. In the pre-quantum statistics of Maxwell and Boltzmann two microstates are already counted as different when only two particles of the same kind have exchanged their mechanical properties, even though the two states are, of course, *empirically* indistinguishable. In other words, the Maxwell— Boltzmann statistics relies on the mere *conceptual* distinguishability, in a mere analogy to a corresponding macroscopic situation where the identity of the particles might be ascertained by following their continuous paths through space and time. In quantum theory, however, two states that differ only in their labels for like particles are considered to be not only *experimentally* undistinguishable but also *conceptually* the same state. For physical theory, this means that all expectation values of observable magnitudes must be invariant under a permutation of particles of the same kind. A simple (though not the only) way of satisfying this requirement is to restrict the admissible functions that describe the state of the particles to two types: symmetric functions that remain unchanged under the exchange of like particles, and antisymmetric functions that change their sign under such an exchange. Philosophically, quantum statistics means that like particles are *in principle* indistinguishable entities. (Lucid philosophical analyses are contained in van Fraassen, 1984. A unified conceptual framework for quantum statistics is developed in Costantini *et al.*, 1983 and related papers. A helpful recent overview is given by Stöckler, 1987).

The important point about this indistinguishability is, of course, that it has testable empirical consequences. One of the most important consequences in the case of antisymmetry, and historically the root of the discovery of indistinguishability, is Pauli's exclusion principle that — if the completeness of the quantum mechanical description is granted — forbids any two electrons to assume the same state and thereby secures the existence of the periodic system of chemical elements. It was again Pauli (1940) who discovered a fundamental theorem about the statistical features of matter and radiation: Under very general and highly plausible assumptions (e.g. the validity of the special theory of relativity for the microworld), like particles with integer spin cannot be in antisymmetric states, and particles with half-integer spin cannot be in symmetric states. If, in agreement with experience, it is moreover assumed that more complicated symmetry properties do not occur in nature, like particles with integer spin, or bosons, are always in symmetric states, those with half-integer spin, or fermions, are always in antisymmetric states. Intuitively, this means that arbitrarily many bosons can behave in the same way, whereas no two fermions can. In other words, particles with integer spin can cluster; they show one possible type of statistical behaviour: Bose-Einstein statistics. Particles with half-integer spin, however, obey another type of statistics: Fermi-Dirac statistics. Intuitively speaking, they push other particles of their kind out of the state they occupy, of which Pauli's original principle is a special case.

Now, what does Pauli's spin-statistics theorem — or the experimental facts that it explains — mean in our context? The spin of elementary particles is an invariant internal property of any such particle like its mass, its lifetime, its size and internal structure (if it has any), or certain quantum numbers. If then all those properties can rightly be interpreted to give us the real structure of those microentities and if we have good reasons to believe in Pauli's theorem, we will have to include the statistical type of the particles in the real constitution of things. In other words: statistical patterns of behaviour do not just figure as a general feature of the microworld, but more specifically they occur in different variants for different kinds of microparticles. Hence, whoever is inclined to suspect that the general feature is somehow due to our epistemological relationship to small dimensions, will find it much harder to maintain this view with respect to quantum statistics.

6. CONCLUSION

I shall now try to summarize the lesson that the recent history of physics teaches us about the function of statistics in science.

It is true that physics, including astronomy, helped to establish statistical methods in science. But the methodological use of statistics in physics has remained very limited. Soon after the discovery of the Gaussian error curve the social sciences took over the leading role in statistics. Around the turn of the century the contemporary discipline of mathematical statistics emerged out of evolutionary biology, not of physics. Whoever believes that statistics is nothing but a methodological tool or a part of the 'logic' of science, will almost entirely miss its impact on physics. Whether or not you share my inclinations toward a realistic interpretation of scientific theories, you will readily admit that statistics in physics makes its more important, and certainly its specific, appearances within the context of *explanatory theories*.

Indeed, we have seen that statistical concepts, e.g. distribution functions or probability amplitudes, are an indispensable element of all physical theories that deal with the constitution of matter. Moreover, I hope to have made clear that this fact cannot be considered as accidental, preliminary or somehow superficial. Already in the mechanical account of equilibrium thermodynamics the statistical explanation of the phenomena is intimately tied to our cognitive interests and needs. It seems doubtful that any non-statistical theoretical approach could ever give us what we want: an explanation of observable properties of macroscopic matter. Those doubts are deepened when we pursue the attempts at an explanation of the overwhelming mass of irreversible physical processes. This suggests the idea that the statistical concepts of the theory may be considered to be an adequate expression of certain overall structures of bulk matter, structures that enable it to pass through irreversible developments and assume stable equilibrium states.

Quantum theory undoubtedly adds more weight to this idea. Short of rather contrived inventions of hidden parameters and processes, we are almost forced to recognize statistical patterns of behaviour as part and parcel of the basic constitution of elementary particles and of atomic or subatomic systems built up from them. In suggesting a realistic interpretation of statistical patterns, I do not want to imply that the nature of those particles and systems is clear or easy to understand, let alone that it can be visualized according to the idealized pattern of classical physics. There is a lot of work left to be done by physicists as well as philosophers. Our understanding of the micro-reality will, I expect, only grow gradually and very slowly. But I cannot find anything impossible or counter-intuitive in the idea that the constitution of reality may require ever new and unfamiliar concepts for its successful description. In particular I do not see why probabilistic tendencies or corresponding statistical patterns should not prominently figure among those concepts. They do this, though mostly in a non-numerical form, in many of our ordinary everyday accounts, e.g. of people and their habits. And what we say, for instance, about the character of a person cannot be visualized either, though it has of course some visualizable consequences. Why should not a similar situation arise in various domains of physical phenomena?

With respect to statistics, the lesson of physics, as I see it, has been aptly described in one sentence by Richard von Mises: "The problems of statistical physics are of greatest interest in our time, since they lead to a revolutionary change of our whole conception of the universe". (1928, p. 219).

Universität Göttingen, FRG

BIBLIOGRAPHY

Bessel, Friedrich Wilhelm. 1838. 'Untersuchungen über die Wahrscheinlichkeit der Beobachtungsfehler'. *Astronomische Nachrichten* **15** 369—404.
Boltzmann, Ludwig. 1872. 'Weitere Studien über das Wärmegleichgewicht unter Gasmolekülen'. In Boltzmann 1968, vol. I, 316—402.
Boltzmann, Ludwig. 1877. 'Über die Beziehung zwischen dem zweiten Hauptsatz der mechanischen Wärmetheorie und der Wahrscheinlichkeitsrechnung respektive den Sätzen über das Wärmegleichgewicht'. In Boltzmann 1968, vol. II, 164—223.
Boltzmann, Ludwig. (1909) 1968. *Wissenschaftliche Abhandlungen*, ed. by F. Hasenöhrl. Leipzig: Barth. Repr. New York: Chelsea.
Brush, Stephen G. 1976. *The Kind of Motion We Call Heat*, 2 vols. Amsterdam: North Holland.
Costantini, Domenico, Maria-Carla Galavotti, and R. Rosa. 1983. 'A Set of "Ground Hypotheses" for Elementary Particle Statistics', *Il Nuovo Cimento* **74B** 151—158.
Eisenhart, Churchill. 1971. 'The Development of the Concept of Best Mean of a Set of Measurements from Antiquity to the Present Day', Unpublished presidential address, American Statistical Association, 131st Annual Meeting, Fort Collins, Colorado.

Gamow, George. 1928. 'Zur Quantentheorie des Atomkerns', *Zeitschrift für Physik* **51** 204—212.

Gauss, Carl Friedrich. 1809. *Theoria motus corporum coelestium in sectionibus conicis solem ambientium.* Contained in: *Werke.* Leipzig/Berlin. 1863—1933.

Gigerenzer, Gerd *et al.* 1989. *The Empire of Chance — How Probability Changed Science and Everyday Life.* Cambridge University Press.

Gurney, R. W. and Condon, E. U. 1929. 'Quantum Mechanics and Radioactive Disintegration'. *Physical Review* **33** 127—140.

Hagen, G. H. L. 1837. *Grundzüge der Wahrscheinlichkeitsrechnung.* Berlin: Dümmler.

Maxwell, James Clerk. 1860. 'Illustrations of the Dynamical Theory of Gases'. In Maxwell 1890, vol. 1: 377—409.

Maxwell, James Clerk. 1878. 'On Boltzmann's Theorem on the Average Distribution of Energy in a System of Material Points'. In: Maxwell 1890, vol. 2: 713—741.

Maxwell, James Clerk. 1890. *Scientific Papers,* ed. by W. D. Niven, 2 Vols. Cambridge: Cambridge University Press.

Pauli, Wolfgang. 1940. 'The Connection between Spin and Statistics'. *Physical Review* **58** 716ff.

Prigogine, Ilya. 1962. *Non-Equilibrium Statistical Mechanics,* New York: Interscience Publishers.

Prigogine, Ilya. 1984. *From Becoming to Being.* 2nd ed. San Francisco: Freeman.

Rutherford, Ernest, and Frederick Soddy. 1903. 'Radioactive Change', *Philosophical Magazine* **5**.

Sheynin, Oscar. 1977. 'Laplace's Theory of Errors', *Archive for History of Exact Sciences* **17** 1—61.

Stöckler Manfred. 1987. *Philosophische Probleme der Elementarteilchenphysik* (Book, to be published).

Van Fraassen, Bas C. 1984. 'The Problem of Indistinguishable Particles'. In: *Science and Reality: Recent Work in the Philosophy of Science — Essays in Honor of Ernan McMullin,* ed. by J. T. Cushing, C. F. Delaney, and G. M. Gutting, University of Notre Dame Press, pp. 153—172.

Von Mises, Richard. 1928. 'Wahrscheinlichkeit, Statistik und Wahrheit'. Quoted from the English translation of the 3rd ed. of 1951 which appeared as *Probability, Statistics, and Truth.* London: Allen and Unwin 1957.

Von Plato, Jan. 1987. 'Probabilistic Physics the Classical Way'. In *The Probabilistic Revolution. Vol. 2: Ideas in the Sciences,* ed. by L. Krüger, G. Gigerenzer, and Mary S. Morgan, Cambridge/Mass.: MIT Press, pp. 379—407.

Young, Thomas. 1819. 'Remarks on the Probability of Error in Physical Observations', Letter to Henry Cater. *Philosophical Transactions of the Royal Society of London* **109**, part I, 70—95.

ABNER SHIMONY

THE THEORY OF NATURAL SELECTION AS
A NULL THEORY

The main thesis of this paper is that the theory of natural selection is a
null theory: it consists of no principles whatever.[1] The theory of
evolution is commonly said to consist of three parts,

1. a theory of heredity,
2. a theory of variation, and
3. a theory of natural selection.

Of these, the first two do indeed have principles, formulated by Darwin
and Wallace phenomenologically, then by de Vries, Morgan, and others
in microbiological terms, and currently with wonderful precision in the
language of molecular biology. There have been many and varied
attempts to formulate a principle of natural selection which would be
coordinate with these principles of heredity and variation, the formulations typically employing the concepts of fitness, adaptation, or differential reproduction rate. The stark thesis that the theory of natural
selection is a null theory conflicts with all of these. I believe, however,
that the thesis is implicit in the insistence by neo-Darwinians upon the
utterly opportunistic character of the evolutionary process and in their
rejection of any general guiding plan in evolution.[2] Metatheoretically,
what the neo-Darwinians are saying is that the evolution of the biosphere (subsequent to the establishment of the genetic code) is
governed by the principles of heredity and variation and by the laws of
physics, and constrained by environmental boundary and initial conditions *but not otherwise*: within these constraints let happen what
happens. I do not wish to be understood as denying that biologists have
said many valuable things about fitness, adaptation, and differential
reproduction rates, but I do deny that their statements have been made
at the level of first principles. When their assertions are correct, they
are at the level of applications of the theory of evolution or at the level
of specialized and conditional formulations; for at these levels, the
injunction 'let happen what happens' can and often does eventuate in
new constraints.

The thesis that the theory of natural selection is a null theory should

*R. Cooke and D. Costantini (eds), Statistics in Science. The Foundations of Statistical Methods in
Biology, Physics and Economics*, 15—26.
© 1990 *Kluwer Academic Publishers*.

not be conflated with the claim that the theory of natural selection is
tautologous (e.g., Waddington (1957), 64—5). There have been sensible
answers to Waddington's and similar arguments. Mills and Beatty
suggest that

fitness be regarded as a complex *dispositional* property of organisms. Roughly speaking
the fitness of an organism is its *propensity* to survive and reproduce in a particularly
specified environment and population. (in Sober (1984), 42).

Since the propensity to survive and reproduce is not identical with the
actual rate of survival and reproduction, the proposition 'the fittest
survive' is not tautological. (See also Brandon in Sober (1984)). This
answer is fortified by proposals of criteria for fitness in terms of design
or engineering or problem-solving, e.g.,

certain morphological, physiological, and behavioral traits should be superior a priori
as designs for living in new environments. These traits confer fitness by an engineer's
criterion of good design, not by the empirical fact of their survival and spread. It got
colder before the woolly mammoth evolved its shaggy coat. (Gould in Sober (1984),
33).

These statements, and many variants of them, are true or approximately
true in a wide range of circumstances. But they do not suffice to
establish 'survival of the fittest' as a non-null biological principle, even
after a suitably sophisticated explication of the terms; and the same is
true for other formulations of the theory of natural selection. The
reason is that the crucial terms in all these formulations — 'fitness',
'adaptation', 'problem', 'propensity', and even 'probability' — cannot be
taken to be well defined without adequate grounding, which is contingent. And the contingencies in which they are well defined are themselves the consequences of evolution.

Part of my point has been made by Lewontin in his polemic against
oversimplified claims for biological adaptation:

Are there really preexistent 'problems' to which the evolution of organisms provides
'solutions'? This is the problem of the *ecological niche*. The niche is a multidimensional
description of all the relations entered into by an organism with the surrounding world.
. . . To maintain that organisms adapt to the environment is to maintain that such
ecological niches exist in the absence of organisms and that evolution consists in filling
these empty and preexistent niches. (in Sober (1984), 237).

Lewontin, as I understand him, is saying that to the extent that niches, problems, solutions, and adaptations have some existential status in the biosphere, and are not just artifacts of theorists, they are themselves the consequences of evolution — that is, of processes which are random except for the general constraints noted above (the laws of heredity and variation and the laws of physics) together with additional constraints that are the products of the previous history of the populations and environments involved. I agree with Lewontin's analysis, and also with analyses which he has made elsewhere of the constraints that result from history — especially the conservatism of the *Bauplan* (Gould and Lewontin in Sober (1984), 265). Nevertheless, his analysis either misses or fails to make explicit much of what I am trying to convey. Because the concepts of niche, problem, and adaptation are difficult to disentangle from teleology, as Lewontin himself points out (in Sober (1984), 235—7), his claims for their historical and contingent character are too much subsumed under his excellent argument against smuggling teleology into the evolutionary process. In order to make my point in full strength, I shall analyze a concept that is more 'neutral' than these, freer from overtones of value and final causation — viz., the concept of probability.

In *The Nature of Selection* (1984a) Sober makes some excellent observations about probability which he then fails to develop sufficiently. He rejects

The idea ... that a probability assignment is justified only if it reflects every bit of known information. This is the Laplacean idea that is independent of the truth of determinism ... this methodological thesis is mistaken because it fails to take into account the different puposes we have in assigning probabilities. ((1984a), 122).

What is lost by the (extended) Laplacean idea is the explanatory utility (133), the theoretical fruitfulness (130), and the possibility to obtain perspective on similarities and differences among single events (134).

I have no objection to what Sober says positively, but I believe that he has omitted something crucial: the way in which nature must cooperate with the scientist in order to render fruitful the methodology of deliberately neglecting the known and unknown differences among systems of a group. Sober offers the following formal probabilistic treatment of grouping:

Suppose we think that all the systems have in common the property C, though they differ with respect to B_1, B_2, ... , B_n (which are mutually exclusive and collectively exhaustive). We wish to assign each system a probability of displaying the property E. Now it may well be that we would end up assigning the different systems different probabilities if we took into account not only their common feature C but also the fact that they differ with respect to $B_1, B_2, \ldots, B_n$. If we nevertheless talk about all the systems as having a single probability, relative to C alone, this probability must be expandable into the following summation:

$$\Pr(E/C) = \sum_{i=1}^{n} \Pr(E/C\&B_i) \Pr(B_i/C).$$

In claiming that the univocal probability expressed '$\Pr(E/C)$' can be explanatory, I am asserting that the summation probability with which it is equivalent can be too. ((1984a), 133).

In order to go beyond formalities, however, it is essential to inquire whether the probabilities $\Pr(E/C\&B_i)$ and $\Pr(B_i/C)$ are well defined as ontic probabilities or propensities, for I assume, as Sober seems to ((1984a), 43—4), that these rather than epistemic probabilities are employed in the probabilistic assertions of physics and biology. No answer can be given to this question without more information concerning the systems involved and the features C and B_i. Fortunately, exemplary cases in which the question has been answered are provided by the theory of dynamical systems (which includes idealized gambling devices and idealized gases behaving in accordance with Newtonian mechanics).

(a) Let C specify a system of N (greater than 2) perfect hard spheres contained in a rectangular box with perfectly reflecting walls, and with total energy $\mathscr{E}$. The temporal evolution of this system can be represented conveniently by the motion of a point on a $6N - 1$ dimensional energy surface $S_{\mathscr{E}}$ in a $6N$ dimensional phase space. The motion is determined by Newtonian mechanics, together with the assumption that the only forces are those of elastic collisions. Sinai proved the great theorem that this system is *ergodic* — that is, the time average $\bar{f}(x)$ for any well-behaved function f, with the initial position x on the surface $S_{\mathscr{E}}$, is equal to the space average $\bar{f}$ over $S_{\mathscr{E}}$, for almost all points x. (Standard works on ergodic theory give the precise specification of 'well behaved' and the definitions of 'space average' and 'almost all', e.g., Arnold and Avez (1968); Sinai's theorem is discussed on 76—9 and 191—3 of that book, and reference is given to the very intricate complete proof). Now let E be a (measurable) subset of $S_{\mathscr{E}}$; it represents in a natural way a property of the system, designated by the same letter E in the passage by Sober. Let f be the characteristic function associated with E: $f(x) = 1$ when $x \in E$, $f(x) = 0$ when $x \notin E$. Finally, let $B_1, \ldots, B_n$ be a partition of S into n disjoint and exhaustive subsets of non-zero measure. We now have the materials available to define the probabilities mentioned by Sober:

$\Pr(E/C \& B_i) =$ the space average of $\overset{*}{f}(x)$ over the set B_i.

(This is a reasonable definition, because it is the average over all the x in B_i of the average proportion of the time in the long run that the system starting at x will spend in E).

$\Pr(B_i/C)$ can be any real number between 0 and 1, subject to the constraint that the sum of these from $i = 1$ to $i = n$ is 1.

(For example, $\Pr(B_i/C)$ may be the measure of B_i, where the measure function appropriate to $S_{\mathscr{E}}$ is used; but it may be something else, depending upon the mechanism for choosing among $B_1, \ldots, B_n$, which has not been specified in the problem as stated).

The arbitrariness of $\Pr(B_i/C)$ causes no trouble, because it trivially follows from Sinai's theorem that $\Pr(E/C \& B_i) = \bar{f}$ for all $i = 1, \ldots, n$. The equation in Sober's passage then yields the definite expression $\bar{f}$ for $\Pr(E/C)$ — which is not at all surprising, in view of ergodicity. Hence, with nature's (and Sinai's) help we have good reason for neglecting the differences among the different systems which have only C in common.

(b) Let C be exactly as before, except that it does not specify the energy $\mathscr{E}$, and let the B_i' be the specification $\mathscr{E} = \mathscr{E}_i$ (entirely different from the B_i of example a). We let the mechanism for choosing among $B_1, \ldots, B_n$ be specified in such a way that a definite value $\Pr(B_i'/C)$ is given — e.g., by a gambling device or a Boltzmann distribution of energies at some temperature. Finally, let E be a subset of the phase space, which may intersect any of the energy surfaces $S_{\mathscr{E}_i}$. We now define

$$\Pr(E/C \& B_i') = \bar{f}_i, \text{ which is the space average of } \overset{*}{f}(x) \text{ over the surface } S_{\mathscr{E}_i}.$$

Now all the probabilities in Sober's equation are well defined. Even though

$$\Pr(E/C) = \sum_{i=1}^{n} \bar{f}_i \Pr(B_i/C)$$

is not a familiar quantity in the theory of dynmaical systems, and even though the $\bar{f}_i$ may differ widely among themselves, $P(E/C)$ may still be very informative about the long run average behavior of the ensemble of systems prepared in the way just described.

Clearly much relaxation of the idealizations of examples a and b can be allowed without complete loss of explanatory utility. There could be variants of example a in which the $\Pr(E/C\&B_i)$ vary, but not too much, or in which those instances of i for which $\Pr(E/C\&B_i)$ differ drastically from the mean are precisely those for which $\Pr(B_i/C)$ are very small. And there could be variants of example b in which the $\Pr(E/C\&B_i)$ differ drastically among themselves but the value of each is only roughly fixed, and likewise for the $\Pr(B_i/C)$: and nevertheless, the uncertainties are sufficiently bounded that something interesting can be said about $P(E/C)$.

A crucial question for the theory of natural selection is whether there is any reason to think that *in general* there are well defined probabilities $\Pr(E/C)$, where E is a typical proposition concerning the reproductive success of a population and C is a typical body of information that biologists provide about the population and its environment. I am very skeptical that a positive answer can be justified. C usually provides a rough characterization of a population in a certain region (for instance, the statistics of its phenotypic varieties), and also an account of the relatively fixed features of the environment, together with some information about the ranges of the variable features (e.g., annual rainfall, mean temperature, numbers of predators of different kinds). An interesting choice of $B_1, \ldots, B_n$ is a partition of the space of variable features of the environment into n mutually exclusive and exhaustive subspaces. Let us tentatively assume that $\Pr(E/C\&B_i)$ is defined for each i from 1 to n. There is no *a priori* and general reason for believing that the $\Pr(E/C\&B_i)$ are fairly close to each other, as in softened variants of example a above, for the viability of the population described by C may be highly unstable with respect to variation of the features of the environment. In other words, it is possible that the population in question behaves like the systems studied in the theory of chaos.[3] Furthermore, the biological situation may not be a softened variant of example b, because the probabilities $\Pr(B_i/C)$ may not be defined at all. At a time of geological instability, the $\Pr(B_i/C)$ may not be defined in principle (as 'unknown probabilities'), much less computable in practice. For these reasons, $\Pr(E/C)$ may not even have a nontrivial interval value, and certainly not a point value. Furthermore, we may have conceded too much in assuming that the $\Pr(E/C \& B_i)$ are well defined if $B_1, \ldots, B_n$ is a coarse partition (as the finitude of n entails). The kind of reasoning which has just been presented leads us to be unsure whether the probability of E upon anything less than a complete specification of the environment and the population is well defined, in which case we have relapsed to the 'Laplacean idea' which Sober sensibly rejected.

Skepticism about the general definability of ontic probabilities $\Pr(E/C)$ in a biological context does not imply a sweeping rejection of properly conditionalized ontic probabilities in evolutionary biology and population genetics. A few remarks will indicate that my scruples leave a vast domain for the responsible use of probability theory in biology.

(1) When selection experiments are performed in a laboratory, a particular B_0 is selected, and hence the uncertainties of the $\Pr(B_i/C)$ are eliminated. If, then, the probabilities of various propositions $E_1, \ldots, E_k$ concerning competing populations (e.g., variant strains of bacteria) are well defined relative to B_0, then these probabilities will be manifested in the actual reproductive performance of these populations. Of course, the word 'manifested' must be used with the usual *caveat* of probability theory, since definite probabilities of single events only determine the probabilities of the possible constitutions of sequences of events; but standard probability theory deals with this complication. With attention to this *caveat*, one can even make a large number of repeated experiments in order to check whether $\Pr(E_i/C \& B_0)$ seems to have a definite value.

(2) In a time interval which is long compared to one generation of the population of interest the environment may as a matter of fact change little, or change in a statistically definite way, thus permitting the probabilities $\Pr(B_i/C)$ to be inferred from observational data. Then $P(E/C)$ may be determined by calculation, if the $\Pr(E/C \& B_i)$ are known, or from observational data about the performance of the population over a long enough period of time to provide a good averaging over the various B_i. It seems to me that the satisfaction of this condition provides the justification for the responsible use of probabilities in population genetics. Whether this kind of justification is also available in an evolutionary study depends upon a comparison of time scales: the time scale over which the environment is statistically stationary must be as long as the time scale over which a nontrivial

evolution of the population of interest takes place. Whether the time scales satisfy this condition is obviously a contingency.

(3) Although the exemplary instances cited above of ontic probability were drawn from the theory of dynamical systems, one should not conclude that the typical mechanisms in those systems (e.g., efficient randomizing mechanisms, as in a container of colliding molecules) are the only ways to account for ontic probabilities. We know that a living cell is an elaborate system of chemical cybernetics, which for the most part operates homeostatically to maintain proper concentrations of sugars, proteins, nucleic acids, lipids, etc. Cybernetic mechanicsms also operate in the integration of multi-celled organisms, and, to a lesser extent, in populations of organisms. It may be that cybernetic mechanisms like cooperation will to some extent mitigate the instability of a population with respect to variation of environmental parameters — an instability which was mentioned above as a possible obstacle to having well defined biological probabilities. If this mitigation occurs, it is a contingency, which cannot be relied upon generally, and it seems to me to be a fascinating problem in a concrete case to find out whether and to what extent it does occur.

After this acknowledgement of some interesting contingencies under which ontic probabilities may be well defined in biology I return to my main theme. There is no hope of supplementing the null theory of natural selection by principles which are formulated in terms of biological probabilities, because these probabilities are only conditionally defined. There is no reason to exclude the possibility that a large part of evolutionary change occurs in situations which are too chaotic for the relevant propositions to have well defined point probabilities or even interval probabilities. Since the concepts which are used in proposed non-null formulations of the theory of natural selection, such as fitness, adaptation, and propensity to reproduce, are essentially probabilistic (see the quotation from Mills and Beatty earlier in the paper), the generality of these formulations is destroyed by the fact that biological probabilities are only conditionally defined.

I can sharpen this criticism by commenting upon some specific proposed formulations. As illustrations I shall consider two principles of Williams (in Sober (1984) 86—7), which she calls 'translations of axioms' (in contrast to 'axioms', which contain some uninterpreted terms):

D3: For each organism, b_1, there is a positive real number, $\phi(b_1)$, which describes its fitness in its environment.

D4: Consider a subcland D_1 of D. If D_1 is superior in fitness to the rest of A for sufficiently many generations (where how many is 'sufficiently many' is determined by how much superior D_1 is and how large D_1 is), then the proportion of D_1 in D will increase during these generations.

Central in her 'intuitive introduction to the concept of fitness' is that

Fitter organisms have a better *chance* of surviving long enough to leave descendants, but a fitter organism does not necessarily leave more descendants than its less fit brother. (*ibid.*, 86).

Since 'chance' is clearly used here synonymously with 'probability', my assessment of D3 is the following. Whatever the details concerning the function ϕ may be, it depends upon the relevant biological probabilities of propositions about b_1 in its environment, where the environment is somehow characterized, perhaps completely but, more likely, only partially; and therefore when these probabilities are not well defined, $\phi(b_1)$ does not have a definite value, contrary to D3. I also suspect, but cannot definitively establish because of the sketchiness of the treatment of ϕ, that when D3 does hold in a case, it does so analytically, as a consequence of the meaning of ϕ and the values of the relevant probabilities, and hence the axiom is superfluous. D4 will almost certainly not hold in every case in which the fitnesses involved are well defined, if 'sufficiently many' is finite, for the reason that Williams herself indicates: that the frequency of 'successes' in a finite sequence of trials need not be equal to or close to the probability of 'success' in each trial. Biologists have recognized deviations of actual frequencies from the underlying probability (assumed to be well defined) to be a nontrivial factor in evolution and have named it 'random genetic drift' (Sober (1984a), 110—3); and this factor is denied if D4 is interpreted literally. If, however, D4 is intended less stringently, as saying that 'the proportion' of D_1 in D will probably increase (i.e., with probability $1 - e$, where e is a small positive number depending upon 'sufficiently many'), then D4 would state nothing more than a simple consequence of probability theory.

My argument can now be summed up. The contingencies in which 'ontic probability', 'degree of fitness', 'degree of adaptation', 'niche', 'problem', etc. are well defined are themselves the products of evolution. Consequently, there is no possibility of formulating a general principle constraining the process of evolution in terms of these concepts. A general theory of natural selection must be independent of contingencies, and the only theory which satisfies this demand is the null theory.

The objection may well be raised that with the dissolution of all non-null general principles of natural selection, the explanatory power of the theory of evolution as a whole will be fatally subverted. *Prima facie*, this objection is strong when directed towards the problem of the precision of adaptation — the exquisite aerodynamical engineering of birds' wings, the chemical engineering of systems of alimentation, the optical engineering of the eye, etc. However, the outlines of an answer should be evident from earlier parts of this paper. The problems which a biological lineage has to 'solve' are not posed *a priori*, but are themselves the outcomes of evolution, as Lewontin correctly insists. But the time scale on which a problem may be fairly well posed in terms of the life strategies of a population — e.g., the problem of flying efficiently, or of discerning prey at a distance — may be very long compared to the time scale over which 'solutions' occur by variation. When these conditions are satisfied, the relative probabilities of reproductive success of the variants may be well enough defined, as interval probabilities if not as point probabilities, for typical probabilistic arguments to apply. It may even be possible to deploy the mighty mathematical fact that

$$\lim_{n \to \infty} (p_1/p_2)^n = 0, \qquad \text{if } p_1 \text{ and } p_2 \text{ are non-negative real numbers and } p_1 < p_2.$$

in order to explain the probable extinction in the long run of a variety that is ever so little less fit than a competitor. To assume that ontic biological probabilities are always well defined, and not merely well defined in contingencies, is therefore overkill for the purpose of explaining precision of adaptation, and biology is underarmed for overkill.

I conclude by addressing the question: if the theory of natural selection is a null theory, how can it be tested? The answer is that *it*, of course, is not tested. What is tested is the proposition that the theory of evolution as characterized at the beginning of the paper — with principles of heredity and variation and constrained by the laws of physics and by environmental contingencies, but not otherwise — is sufficient to explain both the grand features of the development of the biosphere and the phenomena of interest concerning specific biological developments. A non-null formulation of the theory of evolution would presumably assert a similar proposition, with one crucial change: in place of 'and not otherwise constrained' it would propose a constraint. I must leave vague the word 'explain', which is particularly problematic in biology. However, the problems of explicating 'explanation' are the same for both the null and the non-null formulations, and therefore for the question at issue we can be content with the prevailing standards and modes of explanation among evolutionary theoriests: e.g., the rationalization of the fossil record, the rationalization of phylogenetic relations among organisms, consistency with the known data on rate of mutations, consistency with the molecular biological information about mutagenesis, etc. The crux of the matter is whether there is any evidence for an additional constraint which a non-null formulation of a principle of natural selection postulates. The empirical data bearing upon the null and the non-null formulations of the theory of natural selection must somehow — directly or indirectly — reveal the presence or absence of the postulated constraint. My thesis is that no general constraint upon selection has been or will be seen.[4]

Boston University, Boston, MA 02215, U.S.A.

Boston University, Boston, MA 02215, U.S.A.

NOTES

[1] About five years ago I posed to Dr. Hyman Hartman the question of whether the null theory of natural selection occurs in the literature. He replied that Peirce had it in the famous passage,

The Darwinian controversy is, in large part, a question of logic. Mr. Darwin proposed to apply the statistical method to biology. The same thing has been done in a widely different branch of science, the theory of gases. Etc. (1934; 226).

I think I agree with Hartman, but I am not sure, because I cannot tell from this passage or from other writings of Peirce whether he believed the kinetic theory of gases to be free from all principles other than those borrowed from mechanics and mathematics. Did he think that any additional principles are tacitly assumed in the use of probability theory?

The view of Popper (1974), 134—7, that the theory of natural selection is unfalsifiable is not the same as my thesis that it is a null theory, as the last paragraph of my text should make clear.

[2] A spirited exposition of these neo-Darwinian themes is given by Monod (1971), Chapter 2 and 7.
[3] See, for example, Devaney (1986). Numerous references to chaos occur in recent papers in population genetics and evolutionary biology, but I have seen none which suggested a connection between chaos and the thesis that the theory of natural selection is a null theory.
[4] A further argument for my thesis is given in a paper entitled 'The Non-existence of a Principle of Natural Selection', in *Biology and Philosophy* **4**, 253—273 (1989).

BIBLIOGRAPHY

Arnold, V. I. and Avez, A. 1968. *Ergodic Problems of Classical Mechanics*. New York: Benjamin.
Brandon, Robert. 'Adaptation and Evolutionary Theory', *Studies in the History and Philosophy of Science* **9** no. 3. 181—206 (1978). Reprinted in Sober (1984).
Devaney, Robert L. 1986. *An Introduction to Chaotic Dynamical Systems*. Menlo Park, CA: Benjamin-Cummings.
Gould, Stephen Jay. 1977. 'Darwin's Untimely Burial', in S. J. Gould, *Ever Since Darwin*. New York: Norton. Reprinted in Sober (1984).
Gould, Stephen Jay and Lewontin, Richard C. 1978. 'The Spandrels of San Marco and the Panglossian Paradigm: A Critique of the Adaptationist Programme', *Proceedings of the Royal Society of London* **205** 581—598. Reprinted in Sober (1984).
Lewontin, Richard C. 1980. 'Adaptation', *The Encyclopedia Einaudi*. Milan. Reprinted in Sober (1984).
Mills, Susan and Beatty, John, 1979. 'The Propensity Interpretation of Fitness', *Philosophy of Science* **46** 263—286. Reprinted in Sober (1984).
Monod, Jacques. 1971. *Chance and Necessity*. New York: Vintage Books.
Peirce, Charles S. 1935. *Collected Papers*, vol. 5, ed. C. Hartshorne and P. Weiss. Cambridge, MA: Harvard U. Press.
Popper, Karl R. 1974, 'Darwinism as a Metaphysical Research Programme'. In *The Philosophy of Karl Popper*, ed. by P. A. Schilpp. La Salle, IL: Open Court.
Sober, Elliott. 1984, *Conceptual Issues in Evolutionary Biology: An Anthology*. Cambridge, MA: MIT Press.
Sober, Elliott. 1984a. *The Nature of Selection*. Cambridge, MA: MIT Press.
Waddington, C. H. 1957. *The Strategy of the Genes*. London: Allen and Unwin.
Williams, Mary. 1973. 'The Logical Status of Natural Selection and Other Evolutionary Controversies'. In *The Methodological Unity of Science*, ed. by M. Bunge. Dordrecht: Kluwer Acad. Publ. (1984).

MARIA CARLA GALAVOTTI AND GUIDO GAMBETTA

CAUSALITY AND EXOGENEITY IN ECONOMETRIC MODELS

The following pages are intended to show how statistics enters into a relatively young discipline as econometrics. The role of statistics in the classical inference procedures, namely estimation and hypotheses testing, will not be dealt with. This should not be taken to suggest that within econometrics such procedures should not be the object of attention. On the contrary, several authors have stressed the need for extensive use of statistical tests as tools for mis-specification analysis and diagnostic checking. Here we will be pursuing a different task, namely to point out that statistics also plays an explicit role in the earlier phase of model specification. In particular, we want to show how the form of the model — and hence what kind of inferences can be drawn on its basis — depends on the crucial concepts of non-causality and exogeneity defined in terms of conditional distributions of observable variables.

1. TWO APPROACHES

In 1930 a group of prominent economists (R. Frisch, I. Fisher, J. Schumpeter, J. M. Keynes, among the others) founded the International Econometric Society with the aim of transforming economics into an empirical science by combining three tools: economic theory, mathematics and statistics.[1] The new name proposed for economic science conceived in this way was econometrics.

After several decades, we can now see that the primary purpose has not been fully achieved: econometric method is not *the* method, but only *one* of the methods of economics (though perhaps the most frequently utilized). Instead, econometrics has become a new discipline with a restricted scope: the quantification and measurement of economic relations using observed data.

Traditionally econometrics has been concerned above all with the procedure of specifying a model (called the econometric model) intended to bridge the gap between theory and data. A model of this sort represents a compromise between adherence to theory and coherence with data. Secondly, econometrics has been concerned with

R. Cooke and D. Costantini (eds), Statistics in Science. The Foundations of Statistical Methods in Biology, Physics and Economics, 27—40.
© 1990 Kluwer Academic Publishers.

the construction of statistical methods of estimation, testing and fore-casting, specifically devised for application to economic phenomena. In this respect, the methods proposed are often specific for situations in which observable phenomena are measured by means of nonexperi-mental data, i.e. in cases where only one sample can be generated for each situation. Given the deterministic theory supplied by the econo-mist, the econometrician applies his method to give numerical values to the unknown parameters of the theory.

According to the 'classical' approach, this method can be described as follows:

1. to build a simplified model (in general a linear model) of a deter-ministic economic theory;
2. to add to each relationship provided by the theory an 'error term' or a 'residual' (sometimes called a 'disturbance', see Haavelmo[2]);
3. to estimate the unknown parameters of the model on the basis of observable data;
4. to apply a set of statistical tests to assess the fit of the econometric model to the given data;
5. to use the accepted model for purposes of forecasting and evalua-tion of economic policy interventions.

In this perspective econometric method moves from theories to facts, through a series of predetermined steps, which end with an eco-nometric model, taken to represent the theory that has been assumed. As far as it rests upon precise hypotheses suggested by economic theory the model is 'correct' in the sense that it cannot be mis-specified. Indeed, the pre-eminence of theory over data in the specification of models can be considered a central feature of this approach, which der-ives from the work of the founders of econometrics — like R. Frisch and J. Tinbergen[3] — and can be taken as a sort of 'received view'.

On the occasion of the publication of the twenty-fifth volume of *Econometrica* T. Haavelmo summarized the activity of econometrical research following the foundation of the Econometric Society as:

'the most direct, and perhaps the most important, purpose of econometrics has been the measurement of economic parameters that are only loosely specified in general economic theory'. In so doing econometrics has made it possible 'to screen economic theories and to make them more specific by confronting them with economic facts'.[4]

In the mid-Sixties some papers[5] explicitly introduced uncertainty and risk into economic models, thus bringing about a redefinition of several economic concepts in probabilistic terms. This process has taken a long time to complete. It was only in 1979 that a textbook was written explicitly incorporating a stochastic macroeconomic theory.[6] The transition from a deterministic to a probabilistic approach to economic theory has triggered a reconsideration of concepts and methods of econometrics. This has involved, in the first place, some reflection on the link between theory and data within the construction of models.

As a matter of fact, the relation between theoretical information and the formulation of hypotheses guiding model construction had already been the object of discussion, as witnessed, *inter alia*, by the famous debate between T. Koopmans and R. Vining. The problem at stake there was the role of theories in the observation and measurement of economic variables. Strongly convinced of the pre-eminence of theories over data, Koopmans stressed the dependance on theories of the specification of error terms and of underlying assumptions in the construction of models. Moreover, he maintained that the relevance of econometric models to economic policies needs to be justified on the basis of theoretical considerations. According to this point of view theories give the structural framework within which data are to be organized. On the other hand, Vining emphasized the need to take into account the characteristics of data, and stressed the importance of statistics in this respect. He favoured the adoption of a 'statistical point of view' as a 'conception of nature no less useful in the study of economic variation than in the study of physical phenomena'.

According to this point of view statistical methods should not be seen merely as tools for the evaluation of relations postulated by economic theories. Instead, 'probability theory is fundamental as a guide to our understanding of the nature of the phenomena to be studied and not merely as a basis for a theory of the sampling behaviour of estimates of population parameters the characteristics of which have been postulated'.[7] At the time, this debate probably had the drawback of enhancing the idea that a priori theoretical knowledge plays a privileged role in model building, as held by Koopmans. A more recent trend in model building, however, seems to recall Vining's position.

A shift in emphasis from theories to facts is actually the main characteristic of the passage from the 'received view' of econometric method to the alternative approach. This stresses the importance of the structure of observed data, and broadens the scope of econometric modelling into 'the systematic study of economic phenomena using observed data'[8]. In this approach, advocated by a number of authors including D. F. Hendry[9] and A. Spanos, a central role in econometrics is assigned to statistical models specified in terms of observable random variables. On the one hand, statistical models are to be distinguished from theoretical ones, on the other observed data are to be kept separate from theoretical variables. A theoretical model represents a mathematically formulated construct making use of theoretical variables (such as for example demand and supply in relation to intentions and plans of economic agents). A statistical model supplies probability distributions for the observed variables and specifies a sampling model. The validity of a statistical model rests on the probabilistic structure of the observed data, and not on that of the theory.

Theory has of course a crucial influence on the specification of statistical models, since it guides the choice of data to be observed. To a theoretical variable, however, there might correspond several series of observable data. The task of statistical models is precisely that of summarizing all the information provided by observed data, and of organizing such information into a suitable statistical form. A main feature of the approach at hand is the postulation of a 'data generating process' at the basis of econometric modelling, that is of a stochastic process generating the variables taken into consideration. Statistical models can be seen as hypotheses about the structure of this process. As such, they are to be taken as pragmatic tools, and should not be interpreted as 'true' models of reality.

Theoretical questions, such as inference and predictions regarding theory, are dealt with by means of (empirical) econometric models in a proper sense. These are constructed on the basis of statistical models, but have both a statistical and a theoretical meaning. In general an econometric model is obtained through a reparametrization/restriction of a statistical model. In a similar way an estimable model is also obtained. This shows the indispensable function of statistical models in this view. As pointed out by Spanos, prior to any theoretical question the validity of the statistical model chosen has to be ensured. The fundamental importance of statistical models is also linked to the application of testing procedures. As a matter of fact, the approach at hand makes extensive use of statistical tests in order to verify the correct specification of models. This is strongly stressed by D. F.

Hendry, who claims that 'the three golden rules of econometrics are test, test and test'.[10] The crucial role of tests is of course related to the nonexperimental character of econometrics. Where proper randomization procedures cannot be applied, extensive use of tests is unavoidable to guarantee the correct specification of statistical models.

A central feature of this perspective is the fact that — as suggested by the preceding considerations — a plurality of models is admitted. This can be seen as a consequence of a pragmatic view of model building, according to which this is an activity directed towards the achievement of certain (limited) objectives. Different models are taken as suitable for different purposes. For example, statistical models are generally sufficient for making estimates and forecasts, while econometric models are needed for description and explanation. The statistical model which satisfies test criteria, and which is called the Well Defined Estimated Statistical Model represents the basis for the choice between two competing econometric models derived from different theories. This common statistical framework is necessary to give meaning to the choice in case competing models are non-nested, i.e. where neither of two competing models can be taken as a special case of the other.

It is noteworthy that a pluralistic view of this sort is not in itself new, having already been set forth by P. Suppes in his paper 'Models of Data' in 1962.[11] Here the author maintains that a hierarchy of models is needed to connect theories to observable data. This paper is never quoted by econometricians, even though it does contain some of the ideas advanced for example by Hendry and Spanos.

In the epistemological literature, a view like the one upheld by the above mentioned authors has a counterpart in that put forward by B. van Fraassen. According to this author a pragmatic component enters as an essential part both in scientific practice and in its rational reconstruction. In particular, model building is a pragmatic and pluralistic activity and occupies a central place in science and epistemology. In van Fraassen's words: 'scientific activity is one of construction rather than discovery: construction of models that must be adequate to the phenomena, and not discovery of truth concerning the unobservable'.[12]

2. CAUSALITY AND EXOGENEITY

The specification of the econometric model, as we have seen, requires additional assumptions with respect to statistical models. These are in general causal assumptions, largely based on the notion of exogeneity. We now turn to the role played by the concepts of causality and exogeneity in the specification of econometric models. We will see, in particular, how in the passage from the 'classical' to the probabilistic approach to econometrics, together with that of model, the notions of causality and exogeneity were radically reappraised and redefined.

Causality has always played a central role in econometrics, where, in view of the operative character of this discipline, it has traditionally been related to that of manipulability. In short, the notion of causality adopted within econometrics results from a combination of functionalism with a manipulative view. Such a notion rests on the idea that causality is essentially an ingredient of model building. In other words, causality is seen as a category applying to models, and is used to specify certain features of them. This conviction is clearly stated in the works of H. Simon and H. Wold, who probably gave the notion at hand its most perspicuous formulation in the context of the 'received view' of econometric method. '. . . my term 'cause' — says Simon — does not appear anywhere in the mathematical model. It is employed only to talk *about* (denote) a property of the model; hence it occurs in the language we use to talk about the model — a metalanguage'.[13]

The peculiarity of a model that qualifies as causal is then to be identified with the possibility of manipulating the relations fixed by it. This is what distinguishes a causal model from other predictive models: 'if a forecasting model is to be applied for purposes of economic policy in terms of instruments and targets . . .' — says Wold — 'the relations between instruments and targets must be specified as (hypothetical) cause-effect relationships'.[14] One can then say that in econometrics the causal specification of a model rests on the need to use it for operative purposes, and not just for forecasting. Such a pragmatical characterization of causality holds both for the 'classical' and for the probabilistic approach to econometrics, and can be taken as a unifying property of the otherwise different definitions of causality put forward in these contexts.

Let us now briefly recall these definitions. The main feature of Simon-Wold's view, and in general of what we have labelled the 'classical' approach, is that causality is analyzed in a qualitative fashion. The notion of causality is defined on the basis of a priori conditions, usually imposing some restrictions on the parameters of the model.

When such conditions can be justified by an appeal to economic theory, in the sense that they are directly suggested by it, the model is called 'structural'. A structural model is usually interpreted as describing the behaviour of an economic system, according to some theory. This is clearly in agreement with the 'received view' of econometric models that we have previously summarized. Here causality, through the notions of exogeneity and structure, is intimately linked with theoretical hypotheses.

This 'received view' depends in fact on a number of assumptions that several authors, like R. E. Lucas, T. J. Sargent and C. Sims have criticized in various respects.[15] For example, such hypotheses as the constancy of economic parameters over time and the practice of taking exogenous variables as non-random variables, fixed once and for all at the outset, have been shown to be incompatible with the information on which one relies in order to qualify a model as 'structural'. Arguments of this sort have favoured a reconsideration of the notion of causality. This has come together with the adoption of a probabilistic approach.

The best known theory of causality put forward in this connection is that of C. W. J. Granger.[16] Here the definition of causality is based on the information provided by data, arranged in time series. Such a definition conveys the idea that a variable y can be said to 'cause' another variable, say x, if the information provided by the knowledge of y at time t is relevant to the value taken by x at time $t + 1$ (assuming that all considered variables are measured at prespecified time points at constant intervals $t = 1, 2, \ldots$). Causality is then defined in terms of statistical variables and time enters its definition as an integral part. Incidentally, it might be noted that Granger's definition of causality is close to that put forward by P. Suppes.[17] Actually, Suppes' causality represents a more general notion, and Granger's causality can be seen as a particular case of it.

To give an operative import to his notion of causality, and to connect it with that of predictability, Granger works out the complementary notion of non-causality, essentially based on that of conditional independence. Non-causality can be subjected to various tests procedures, which specify its operational meaning. Granger's causality is in the spirit of Sims' criticism, and rests on the idea that causality is to be based in the first place on data, not on theory. As a matter of fact, Granger claims that theoretical knowledge has no role at all in his theory. However, to justify this claim he embodies in it a sort of requirement of total evidence, imposing that 'total knowledge of all relevant distribution functions' be taken into account. Assertions of causality are then made conditional on knowledge of the past history of the universe, which should also include information on which variables are to be taken as exogenous and which as endogenous. In practice, however, the requirement can never be fulfilled, as Granger himself seems to recognize when he puts forward a notion of 'causality in mean' not subjected to it, for operative purposes.

Granger's theory has the merit of having introduced the notion of probabilistic causality into econometrics, but at the same time this notion is not sufficient to allow specification of models, because it does not include assumptions about the parameters, and it is then model independent. Recently, R. F. Engle, D. F. Hendry and J.-F. Richard proposed a more comprehensive theory (also set forth in terms of observable variables and conditional probabilities), which can be seen as a generalization of Granger's causality. In this theory, exogeneity is not a distinctive characteristic of a variable as in the classical view (for example policy variables under direct intervention of the authorities). Rather, it is a property that depends on the chosen parametrization of the model and on the use for which the model has been constructed. As a consequence, it is not possible to distinguish endogenous from exogenous variables before the probability distribution of observable variables is considered. In other words, the distinction is not fully definable on a priori grounds, but is to some extent a testable proposition inside a specific model.

Engle, Hendry and Richard define three kinds of exogeneity, characterized by different strengths. A variable z_t of a certain model is said to be *weakly* exogenous with respect to the estimation of a set of parameters λ if the inference about λ conditional on z_t does not involve any loss of information. When a weakly exogenous variable z_t can also be shown to be non-caused (in Granger's sense) by any of the endogenous variables of the model, z_t is said to be *strongly* exogenous. In addition to weak and strong exogeneity the authors define a notion of *super* exogeneity. This applies to the case in which the mechanism generating z_t is allowed to change. A variable z_t is said to be super exogenous in a certain model if it is weakly exogenous, and in addition all the parameters of the model are invariant with respect to changes in z_t, in a given period of time. The changes in the variable defined as super exogenous can take place for a variety of reasons, such as 'changing tastes, technology, or institutions such as government policy making'.[18] In view of this, the notion of super exogeneity can be a useful tool for analyzing the effects of alternative economic policies (provided that they do not involve changes in structural parameters). The notion of super exogeneity relates to a structural property of the model, and can be taken as structural exogeneity. By this means, then, Engle, Hendry and Richard bring the notion of causality back to that of structure, in a probabilistic framework.

Up to now we have been concerned with the construction of models. Of course models must be used for various purposes. In view of this, the notion of structure is not absolute, and does not result from a priori theoretical knowledge. On the contrary, structure is defined in a relativistic way, based on manipulability. Such a notion of structure is to be found in the works of authors like Sims and Hendry, and is largely inspired by the earlier work of L. Hurwicz. This author sets forth a pragmatic view according to which structure is only definable with reference to the set of consequences produced by an intervention on certain objects, whose behaviour is described by the model. The concept of structure is then taken to be relative to a set of possible modifications, and 'represents *not a property of the material system under observation*, but rather a property of the anticipations of those asking for predictions concerning the state of the system'.[19]

Analogously, Sims defines as "structural" a model which is invariant under a certain set of hypothetical interventions. As the author points out, in a perspective like this whether a model is structural depends on the use to which it is to be put — on what class of interventions is being considered.[20] So we cannot decide whether a model is structural (whether it describes a structure) by examining the form of the system of equations, as in the classic view of Simon and Wold. Rather, the property of being structural is related to the application of the system to the world. Such a property can be tested by comparing the predictions as to the effects of an intervention, made on the basis of a certain system, with the observed results of such intervention. Since the model is stochastic there is of course no guarantee that we can obtain the same result repeating the intervention, so that the structure is not a deterministic framework to predict the behaviour of economic agents.

One can then see that — as already noted — the notion of manipulability is an essential component of that of causality, both in a deterministic and in a probabilistic context. In either case, when one talks about causal relations referring to econometric models, one talks about relations that can be manipulated. Moreover, one talks about relations that belong to a structural model. In this connection, the adoption of a probabilistic approach to econometrics has created peculiar problems, partly connected to the nonexperimental character of this discipline.[21] In general, anyway, econometric model building requires a relativistic, context dependent notion of structure, which is itself based on that of manipulability.

APPENDIX

In order to show the meaning of the different types of models involved in econometrics, take the following example. This simple four equation system is a typical representation of a *theoretical model*.

(1) $y_t^D = \delta x_t + \varepsilon_{1t}$ (demand equation)

(2) $y_t^S = \beta x_t^e + \varepsilon_{2t}$ (supply equation)

(3) $y_t - y_{t-1} = \alpha(y_t^D - y_{t-1}) + \varepsilon_{3t}$ (adjustment equation)

(4) $x_t^e = x_t + \mu(x_t - x_{t-1}) + \varepsilon_{4t}$ (expectations equation).

In equation (1) the quantity y_t^D of a commodity desired by the consumers (an unobservable variable) is a (linear) function of the observed price of that commodity, x_t. In equation (2) the quantity supplied by the producers y_t^S is a function of the expected price x_t^e (yet an unobservable variable). Equation (3) expresses the fact that in each period the change in the observed quantity $(y_t - y_{t-1})$ usually is not sufficient (as long as $0 < \alpha < 1$) to fill the gap between the previous level y_{t-1} and the desired level y_y^D, so that the observed quantity partially adjusts to the desired level. In the last equation it is assumed that expectations of the price x_t^e are formed in an extrapolative manner.

The ε_{it} are normally distributed random variables with zero means and finite variances σ_{ii} while the covariances σ_{ij} are supposed to be zero.

The estimation of the theoretical parameters of interest (which have a precise economic interpretation) is impossible given the presence in the model of three unobservable variables. Solving the model for the observable variables y_t and x_t we obtain the so called *estimable model*.

This is done by substituting equation (1) in equation (3), equation (4) in equation (2), and assuming that there is no lag between the supplied and observed quantity $y_t^s = y_t$:

$$y_t = \beta(x_t + \mu x_t - \mu x_{t-1} + \varepsilon_{4t}) + \varepsilon_{2t}$$

$$y_t = y_{t-1} + \alpha(\delta x_t + \varepsilon_{2t} - y_{t-1}) + \varepsilon_{3t}.$$

Solving for x_t in the first equation we obtain

$$x_t = b_1 y_t + b_2 x_{t-1} + u_{2t}$$

$$y_t = a_1 x_t + a_2 y_{t-1} + u_{1t}$$

where

$$a_1 = \alpha\delta \qquad a_2 = 1 - \alpha$$

$$b_1 = \frac{1}{\beta(1 + \mu)} \qquad b_2 = \frac{\mu}{1 + \mu}$$

$$u_{1t} = \mu\varepsilon_{1t} + \varepsilon_{3t} \qquad u_{2t} = \frac{1}{1 + \mu}\,\varepsilon_{4t} + \frac{1}{\beta(1 + \mu)}\,\varepsilon_{2t}$$

However there is no guarantee that this model fits well the observed data. Suppose for example that a good description of time series observations is a bivariate autoregressive *statistical model*.

$$y_t = \pi_{11}y_{t-1} + \pi_{12}x_{t-1} + V_{1t}$$

$$x_t = \pi_{21}y_{t-1} + \pi_{22}x_{t-1} + V_{2t}$$

where π_{ij} are the statistical parameters of interest which usually do not have a direct economic interpretation. This model is sometimes called the reduced form.

Since the variables y_t and x_t are linear combinations of variables ε_{it} which have been assumed to be normally distributed, they are normally distributed too. Consequently, the equations of the statistical model can be considered the mean (expected) values of the joint distribution of y_t, x_t given y_{t-1}, x_{t-1} and the parameter vector π:

$$D(y_t, x_t | y_{t-1}, x_{t-1}, \pi) =$$

By means of the factorization of the joint distribution $D(y_t, x_t)$ in a conditional distribution $D(y_t|x_t)$ and in a marginal distribution $D(x_t)$ we obtain:

$$D(y_t, x_t|y_{t-1}, x_{t-1}, \pi) =$$
$$= D(y_t|x_t, y_{t-1}, x_{t-1}\pi^*). D(x_t|y_{t-1}, x_{t-1}, \pi^{**}).$$

In this context, x_t is said to be weakly exogenous if the parameters of interest are functions of π^* alone and if π^* and π^{**} are variation free. Moreover y_t does not cause (in Granger sense) x_t if

$$D(x_t|y_{t-1}, x_{t-1}, \pi^{**}) = D(x_t|x_{t-1}, \pi^{**}).$$

Finally, if both conditions are satisfied x_t is said to be strongly exogenous for the parameters of interest. Since these properties are empirically testable, we can find the particular model which is consistent with the statistical description of data.

There are many models which have the same reduced form of the original estimable model. One could be the following.

$$y_t = a_1 x_t + a_2(x_{t-1} - y_{t-1}) + e_{1t}$$
$$x_t = a_3 y_{t-1} + a_4 x_{t-1} + e_{2t}.$$

Such a model however has a different behavioural interpretation. Note that x_t is weakly exogenous in this model while it was not weakly exogenous in the previous one. So the different parametrization entails different properties of x_t variables and a different interpretation of the behaviour of economic agents. For details see the paper by Engle, Hendry and Richard quoted in the bibliography.

Dipartimento di Filosofia, Bologna, Italy
Dipartimento di Scienze Economiche, Bologna, Italy

NOTES

[1] Some historical remarks regarding the foundation of the International Econometric Society are to be found in Frisch (1969).
[2] See Haavelmo (1944).
[3] See Frisch (1969) and Tinbergen (1969).
[4] Haavelmo (1958), pp. 351–352.
[5] See for example Arrow (1964).

[6] See Sargent (1979).
[7] R. Vining, 'Koopmans on the Choice of Variables to be Studied and of Methods of Measurement' in Vining and Koopmans (1949), p. 85. See also Koopmans (1947). Some remarks on the debate between Vining and Koopmans are to be found in Morgan (1987).
[8] Spanos (1986), p. 3.
[9] See Hendry (1980), (1983), (1985).
[10] Hendry (1980), p.47.
[11] See Suppes (1962).
[12] van Fraassen (1980), p. 5.
[13] Simon (1955), p. 194. See also Simon (1953) and (1954).
[14] Wold (1969a), p. 376. See also Wold (1954), (1969b) and (1973).
[15] See Lucas (1976) and (1981), Sargent and Sims (1977), Sims (1980) and (1982).
[16] See Granger (1980).
[17] See Suppes (1970).
[18] See Engle, Hendry and Richard (1983), p. 283.
[19] Hurwicz (1962), p. 238.
[20] Sims (1982), p. 332.
[21] See Pratt and Schlaifer (1984).

REFERENCES

Arrow, K. 1964. 'The Role of Securities in the Optimal Allocation of Risk Bearing', *Review of Economic Studies* **31** 91—96.

Engle, R. F., Hendry, D. F. and Richard, J.-F. 1983. 'Exogeneity', *Econometrica* **51** 277—304.

Frisch, R. 1969. 'From Utopian Theory to Practical Applications: the Case of Econometrics', Nobel Lecture.

Granger, C. W. J. 1980. 'Testing for Causality: a Personal Viewpoint', *Journal of Economic Dynamics and Control* **2** 329—352.

Haavelmo, T. 1944. 'The Probability Approach to Econometrics', Supplement to *Econometrica*, 12.

Haavelmo, T. 1958. 'The Role of the Econometrician in the Advancement of Economic Theory', *Econometrica* **26** 351—357.

Hendry, D. F. 1980. 'Econometrics. Alchemy or Science?', *Economica* **47** 387—406.

Hendry, D. F. 1983. 'Econometric Modelling: the Consumption Function in Retrospect', *Scottish Journal of Political Economy* **30** 193—220.

Hendry, D. F. 1985. 'Monetary Economic Myth and Econometric Policy', *Oxford Review of Economic Policy* **1** 72—84.

Hurwicz, L. 1962. 'On the Structural Form of Interdependent Systems' in E. Nagel, P. Suppes, A. Tarski (eds), *Logic, Methodology and Philosophy of Science*, Stanford Univ. Press: Stanford, pp. 232—239.

Koopmans, T. 1947. 'Measurement without Theory', *The Review of Economics and Statistics* **29** 161—172.

Lucas, R. E. 1976. 'Econometric Policy Evaluation: a Critique', in K. Brunner and A.

H. Meltzer (eds), *The Phillips Curve and Labor Markets*, Carnegie-Rochester Conference, Series on Public Policy, n. 1, North-Holland: Amsterdam. Also in Lucas (1981).

Lucas, R. E. 1981. *Studies in Business-Cycle Theory*, the MIT Press: Cambridge, Mass.

Morgan, M. S. 1987. 'Statistics without Probability and Haavelmo's Revolution in Econometrics', in L. Krüger, G. Gigerenzer and M. S. Morgan (eds), *The Probabilistic Revolution*, Vol. 2, The MIT Press: Cambridge, Mass., pp. 171—197.

Pratt, J. W. and Schlaifer, R. 1984. 'On the Nature and Discovery of Structure', *Journal of the American Statistical Association* **79** 9—21.

Sargent, T. J. 1979. *Macroeconomic Theory*, Academic Press: New York, San Francisco, London.

Sargent, T. J. and Sims, C. A. 1977. 'Business Cycle Modelling without Pretending to Have too much a priori Economic Theory', in C. A. Sims (ed.), *New Methods in Business Cycle Research*, Minneapolis: Federal Reserve Bank of Minneapolis.

Simon, H. 1953. 'Causal Ordering and Identifiability', in W. C. Hood and T. C. Koopmans (eds.), *Studies in Econometric Method*, Wiley: New York, pp. 49—74. Also in Simon (1977).

Simon, H. 1954. 'Spurious Correlation: a Causal Interpretation', *Journal of the American Statistical Association* **49** 467—479. Also in Simon (1977).

Simon, H. 1955. 'Causality and Econometrics: a Comment', *Econometrica* **23** 193—195.

Simon, H. (1977), *Models of Discovery*, Reidel: Dordrecht.

Sims, C. A. 1977. 'Exogeneity and Causal Ordering in Macroeconomic Models', in C. A. Sims (ed.), *New Methods in Business Cycle Research*, Minneapolis: Federal Reserve Bank of Minneapolis, pp. 23—43.

Sims, C. A. 1980. 'Macroeconomics and Reality', *Econometrica* **48** 1—48.

Sims, C. A. 1982. 'Scientific Standards in Econometric Modelling', in M. Hazewinkel and A. H. G. Rinnooy Kan (eds), *Current Developments in the Interface: Economics, Econometrics, Mathematics*, Reidel: Dordrecht, pp. 317—340.

Spanos, A. 1986. *Statistical Foundations of Econometric Modelling*, Cambridge Univ. Press, Cambridge.

Suppes, P. 1962. 'Models of Data', in E. Nagel, P. Suppes, A. Tarski (eds), *Logic, Methodology and Philosophy of Science*, Stanford Univ. Press, Stanford, pp. 252—261.

Suppes, P. 1970. *A Probabilistic Theory of Causality*, North-Holland, Amsterdam.

Tinbergen, J. 1969. 'The Use of Models: Experience and Prospects', Nobel Lecture.

van Fraassen, B. C. 1980. *The Scientific Image*, Oxford Univ. Press, Oxford.

Vining, R. and Koopmans, T. 1949. 'Methodological Issues in Quantitative Economics', *The Review of Economics and Statistics* **31** 77—94.

Wold, H. 1954. 'Causality and Econometrics', *Econometrica* **22** 162—177.

Wold, H. 1969a. 'Econometrics as Pioneering in Nonexperimental Model Building', *Econometrica* **37** 369—381.

Wold, H. 1969b. 'Mergers of Economics and Philosophy of Science', *Synthese* **20** 427—482.

Wold, H. 1973. 'Cause-Effect Relationships: Operative Aspects', in P. Suppes, L. Henkin, A. Joja, Gr. C. Moisil (eds), *Logic, Methodology and Philosophy of Science IV*, North-Holland: Amsterdam-London, pp. 789—801.

ROGER M. COOKE

STATISTICS IN EXPERT RESOLUTION: A THEORY OF WEIGHTS FOR COMBINING EXPERT OPINION

INTRODUCTION

In many fields, including risk analysis, reliability, policy analysis, forecasting etc., expert opinion in the form of quantitative subjective probability assessments is being increasingly used when objective statistical data is lacking. This raises new problems for statistitians and probabilists, as models for extracting and analysing data from this new source must be developed. It is not surprising that most models to date have proceeded from the Bayesian standpoint, and intend to support the analyst in 'updating' his prior opinion with the probabilistic assessments from one or more experts. Many examples from this school can be found in (Clarotti and Lindley, 1988). Although there are countless ad hoc applications of expert opinion, applications of models for using expert opinion are harder to find. Good examples can be found in Apostolakis (1988).

This paper describes a 'classical' approach to expert opinion which was developed under contract with the Dutch Ministery of Environment (Cooke, 1989, Cooke *et al.*, 1989). The approach is currently being developed by the European Space Agency (Preyssl and Cooke, 1989). Applications are described in (Bohla *et al.*, 1988; Cooke, 1988, 1990). This approach is anchored in the traditional theory of proper scoring rules which has been proposed for 'evaluating' experts (Winkler, 1969; De Groot and Fienberg, 1986). However, the specific requirements of weighting, as opposed to merely evaluating, motivate an extension of the traditional notion of a scoring rule. The extension hinges on the notion of a scoring rule for average probabilities.

The model described here differs from the Bayesian approaches in that expert probability assessors are treated as classical statistical hypotheses. A 'decision maker's' distribution is formed by taking an optimal weighted average of the experts' distributions. The weights are determined with the help of two important concepts:

— Fisher — type significance tests
— entropy.

R. Cooke and D. Costantini (eds), Statistics in Science. The Foundations of Statistical Methods in Biology, Physics and Economics, 41—72.

The significance tests reflect the degree to which experts' past assessments have 'corresponded with reality' in a sense defined in the following. It turns out that performance on such significance tests is closely related to the 'calibration of subjective probabilities', and in fact, these tests provide a new quantitative measure of calibration which has important advantages relative to the measures currently in use. Entropy measures the degree of (lack of) information in experts' assessments. Experts receiving the greatest weight in this model are those with the best 'calibration' and the lowest entropy (highest information content).

The importance of calibration and entropy in connection with combining expert opinion emerges from a recent psychometric experiment (Cooke et al., 1988) in which the following results were found. On items relating to their field of expertise, experienced operators of sophisticated technical systems are significantly better, as a group, than inexperienced operators with respect to both calibration and entropy. On general knowledge items no significant difference was found. This suggests that good calibration and low entropy both correspond to features which we associate with 'expertise'. However, within each group there was a significant negative correlation between good calibration and low entropy. A 'truly good expert' should be good with respect to both. There was one such expert for the technical items (from a group of 34) and no such expert for the general knowledge items. This indicates the importance of a system of weights which is sensitive to both calibration and entropy.

Section 1 discusses the background literature on weights and scoring rules. Section 2 develops the notion of scoring rules for average probabilities, whose asymptotic properties are discussed in Section 3. Section 4 presents a theory of weights. A final section is devoted to heuristics. Proofs are included in an appendix.

1. BACKGROUND WEIGHTED COMBINATIONS OF EXPERT OPINION

Suppose we have an uncertain quantity and an assessed distribution over the range of this quantity from some expert. A scoring rule is simply a function of the assessed distribution and the observed value of the quantity. A scoring rule is called (*strictly*) *proper* if the assessor's expected score is maximized when (and only when) his assessment represents his true opinion (if the score is negatively sensed, then 'maximized' should be replaced by 'minimized'). Important contributions to the theory of proper scoring rules can be fouond in Shuford *et al.* (1965), Winkler and Murphy (1968), Stael von Holstein (1970), Matheson and Winkler (1976), Friedman (1983), Bayarri and De Groot (1987), and Savage (1971). The latter reference provides the most comprehensive discussion of propriety (the property of being proper). De Groot and Fienberg (1983, 1986) and Winkler (1986) discuss 'goodness' of probability appraisers in relation to scoring rules.

Scoring rules were originally intended as a tool for *eliciting* subjective probabilities. If the scoring rule is strictly proper, then it is in the subject's own interest to state his true opinion (at least if the subject's utility function is linear in the scoring variable, as Winkler (1969) observed). On the other hand, if probabilities are elicited under an improper scoring rule, then (assuming linear utility) the subject is being encouraged to state an opinion at variance with his true opinion.

The theory of weights is concerned with defining a set of positive normalized coefficients for combining probability distributions from a set of assessors for an uncertain quantity. Roberts (1965) showed how Bayes' theorem could be used to update an arbitrary set of initial weights on the basis of the observed values of other uncertain quantities. When a given value of an uncertain quantity is observed, the weight assigned to an expert adviser assessing that quantity changes to a new weight which is proportional to the old weight and to the assessed probability of the observed value.

Winkler (1968) discusses several ways of arriving at a 'consensus distribution' using weights in a Bayesian context. De Groot (1974) (also Wagner and Lehrer (1981)) proposes an elegant theory in which experts weigh each other iteratively and (under certain conditons) tend toward a set of equilibrium weights. Morris (1977) combines expert assessments into a 'composite expert' via a 'joint calibration function'. His theory places very strong restrictions on the decision maker's assessment of the experts, and is quite forbidding computationally. Winkler (1968) suggests using scoring rules to determine weights, without however making a concrete proposal. No concrete proposal for weights based on scoring rules has been made to date.

The main problem in developing a theory of weights for combining expert probability assessments was first signalled by Winkler (1969). If the theory of deriving weights is known to the experts beforehand (as it should be), they will experience their weights as scores, and will try to be as 'heavy' as possible. Winkler showed that Roberts' theory encourages experts to report probability 1 for the outcome for which their subjective probability is highest. Hence Roberts' theory of weights is grossly improper and encourages experts to give extremely overconfident assessments. De Groot and Bayarri (1987) show that the situation improves somewhat if the experts encorporate beliefs regarding the assessments of other experts with whom they are to be weighed. De Groot's 1974 theory must also be considered improper.

On the other hand, the traditional proper scoring rules raise problems of their own when considered as weights. Consider an uncertain quantity with outcomes $i = 1, \ldots n$. Let $p = p_1, \ldots p_n$ be a probabiltiy vector for these outcomes, and let $R(p, i)$ be the score for assessment p upon observing i. The best known strictly proper scoring rules are (Winkler and Murphy, 1968):

$$R(p, i) = 2p_i - \sum p_j^2 \qquad \text{(quadratic scoring rule)}$$

$$R(p, i) = p_i/(\sum p_j^2)^{1/2} \qquad \text{(spherical scoring rule)}$$

$$R(p, i) = \ln(p_i) \qquad \text{(logarithmic scoring rule)}.$$

Under the quadratic and spherical scoring rules the scores depend on the assessments of outcomes which did not in fact occur (for n greater than 2). A scoring rule has been called *relevant* if the scores depend only on the assessments of the observed outcomes. Following Winkler (1969), relevance is felt to be an important property for a theory of weights.

Shuford et al. (1966) prove the surprising result that the logarithmic score (plus a constant) is the only strictly proper scoring rule which is relevant in this sense, when n is greater than 2 (MacCarthy (1956), cited in Savage (1971), attributes this insight to Andrew Gleason).

Hence, if a system of weights is to be relevant and strictly proper, the choice is determined up to the logarithmic scoring rule plus a constant. However, the logarithmic rule has its drawbacks. First of all, its values range from 0 to $-\infty$, and it is positively sensed (higher values are better). In order to get positive, positively sensed weights we must make the weights proportional to $K + R$ for some large constant K. For sufficiently small p_i, however, the weight becomes negative. This may seem like a mere theoretical inconvenience; however, risk analysis, a potential area of application, typically deals with low probabilities.

Indeed, failure probabilities as low as 1.51×10^{-79} have been calculated (Honglin and Duo (1985)).

Friedman (1983) advances several further arguments against the logarithmic score. He contends that it unduly penalizes underestimating small probabilities, and is not 'effective' with respect to any metric, where a score is said to be effective with respect to a metric if reducing the distance (in the sense of the metric) between the elicited and the believed distributions increases the expected score.

De Groot and Fienberg (1986) give an additive decomposition of proper scoring rules into a 'calibration term' and a 'refinement term'. This generalizes an earlier result of Murphy (1973), who speaks of 'resolution' instead of 'refinement' (as 'refinement' applies only for well calibrated assessors, Murphy's term is retained here). The following formulation of their result provides a useful perspective for the ensuing discussion.

Consider N uncertain quantities with identical, finite ranges (e.g. 'win', 'lose' and 'draw'). An expert assigns each quantity to one of B 'probability bins'. Each bin is associated with a probability vector $p_{i_{..}}$ over the range of possible outcomes, $i = 1, \ldots B$ (the notation $p_{i_{..}}$ indicates that $p_{i_{..}}$ is a vector with a component for each possible outcome). The vector $p_{i_{..}}$ is assumed to agree with the expert's distribution of all items in the i-th probability bin. Let $s_{i_{..}}$ be the sample distribution for the items in the i-th bin, and let n_i denote the number of items placed in the i-th bin. Let R be a strictly proper scoring rule, $\mathbf{p} = p_{1_{..}}, \ldots p_{B_{..}}, \mathbf{s} = s_{1_{..}}, \ldots s_{B_{..}}, \mathbf{n} = n_1, \ldots n_B$, and let $R(\mathbf{p}, \mathbf{n}, \mathbf{s})$ denote the total score gotten by summing the scores for all uncertain quantities. If t and r are distributions over the possible outcomes, let $E_t(R(r)) = \Sigma \, t_i R(r, i)$ denote the expected score for one variable with distribution t when the elicited distribution is r. De Groot and Fienberg's decomposition can then be expressed as:

$$(1) \qquad R(\mathbf{p}, \mathbf{n}, \mathbf{s}) = \sum_{i=1}^{B} n_i (E_{s_{i_{..}}} R(p_{i_{..}}) - E_{s_{i_{..}}} R(s_{i_{..}})) + \sum_{i=1}^{B} n_i E_{s_{i_{..}}} R(s_{i_{..}}).$$

$$\underbrace{\hspace{4cm}}_{\text{Calibration term}} \qquad \underbrace{\hspace{3cm}}_{\text{Resolution term}}$$

Note that the right hand side *is* a random variable, despite appearances, as the sample distributions with respect to which the expectations are taken, as random variables. The first term under the summation is the 'calibration term' and the second is the 'resolution term'. Since R is strictly proper, the calibration term equals zero if and only if $p_{i.} = s_{i.}$ for all i. Note that the choice of calibration and resolution terms is completely determined by R. The calibration term in and of itself is not strictly proper.

It is interesting to examine this decomposition for the special case that R is the logarithmic scoring rule. Let A denote the finite set of possible outcomes, and assume that each outcome has positive probability under the distributions p_i over A, $i = 1, \ldots B$. The decomposition (1) of the logarithmic score becomes:

$$(2) \qquad R(\mathbf{p}, \mathbf{n}, \mathbf{s}) = -\sum_{i=1}^{B} [(n_i I(s_{i.}, p_{i.}) + n_i H(s_{i.})].$$

The resolution term is simply the entropy in the joint sample distributions, when these are considered independent. Calibration is measured by the relative information for each bin, weighted by the number of items in each bin. (2) will emerge as a special case of scoring rules for average probabilities. The calibration term by itself will be shown strictly proper, and it will prove possible to assign a multiplicative entropy penalty when entropy is associated with the assessed distributions (propriety in the latter case is asymptotic).

The principal disadvantage in using a scoring variable like (1) or (2) to derive weights (which De Groot and Fienberg do *not* propose) is the following. The resulting scores cannot be meaningfully interpreted without knowing the number of quantities involved and their overall sample distribution.

For example, suppose we score two experts assessing different sets of quantities with (2). Suppose the first expert assesses only one quantity, assigns one of the possible outcomes probability 1, and suppose this outcome is observed. His score is then maximal, namely zero. Suppose the second expert does the same, but for 1000 quantities. His score will also be zero. On the basis of their respective scores, we cannot distinguish the performance of these two experts, and would have to assign them equal weights. Intuitively, however, the second expert should receive a greater weight, as his performance is more convincing (the first expert might have gotten lucky). Dividing R by the number N of uncertain quantities (which De Groot and Fienberg in fact do) would not help.

The point is this: A scoring rule, being a function of the values of uncertain quantities, is a random variable, and interpreting the values of the score requires knowledge of the score's distribution.

Moreover, if S denotes the total sample distribution, then the maximal value of the resolution term in (2) is $NH(S)$. The resolution terms for different sets of uncertain quantities with different sample distributions therefore cannot be compared.

In practice we shall often want to pool the advice of different experts who have assessed different quantities in the past. In light of the above remarks, there would be no way of combining experts' opinions via scores derived from different sets of test variables. Even if the scores did pertain to the same variables, it would be impossible to assess the importance of the differences in scores without some knowledge of the distribution of the scoring variable.

The theory developed below involves scoring variables which are not gotten by summing scores for individual variables. This generalization provides considerably more latitude in choosing (relevant) proper scoring rules. We shall find that the calibration term in (2) is strictly proper in an appropriate sense of the word, and has a known distribution under common assumptions. Weights can then be derived on the basis of the significance level of the calibration term, and these weights are shown to be asymptotically strictly proper. Similar remarks apply to the entropy penalty $\Sigma\, n_i H(s_{i.})$.

It is significant and perhaps surprising that a strictly proper calibration score exists whose sample distribution under customary assumptions is known. The theory presented below generates a large class of strictly proper scores, but only one has been found with this property.

2. PROPER SCORING RULES FOR AVERAGE PROBABILITIES

In this section we develop a theory of strictly proper scoring rules for average probabilities and introduce a useful representation theorem. Proofs of non-trivial results are found in the appendix. $(\Omega,\, F)$ will denote an arbitrary measurable space. It is assumed that all probability measures are countably additive and that all random variables on Ω are F measurable. $\mathbf{R}$ and $\mathbf{N}$ denote the real numbers and the integers respectively. For $A \in F$, 1_A denotes the indicator function of A. The following notation will be adopted:

$0 = \{o_1, \ldots o_m\}$:   set of outcomes;

$X = X_1, \ldots$: $X: \Omega \to 0^\infty$; X is F-measurable.

$M(0)$:   set of non-degenerate probability vectors over 0; $p \in$ $M(0)$, $p = p_1 \ldots p_m$, $\Sigma \, p_i = 1$; $p_i > 0$, $i = 1, \ldots m$.

$k_i^{(N)}$ $= \displaystyle\sum_{j=1}^{N} 1_{\{X_j = o_i\}}.$

$s_i^{(N)}$ $= k_i^{(N)}/N$; $i = 1, \ldots m$;

$s^{(N)}$ $= (s_1^{(N)}, \ldots s_m^{(N)})$.

$M(X)$ set of non-degenerate distributions for X: if $P \in$ $M(X)$, then $0 < P(s^{(N)})$, for all $s^{(N)}$; we assume that if $P \in M(X)$, then P is defined on F. The X_i need not be independent under P.

Q assessor's probability for X; $Q \in M(X)$.

$q_i^{(N)}$ $= (1/N) \displaystyle\sum_{j=1}^{N} Q(X_j = o_i)$;

$q^{(N)}$ $= (q_1^{(N)}, \ldots q_m^{(N)})$; $q^{(N)} \in M(0)$.

We suppose that an assessor is asked to state his/her average over $X_1, \ldots X_N$ of the probabilities of occurrence for the outcomes $\{o_1, \ldots o_m\}$. This is equivalent to asking for the expected relative frequencies of occurrences of these outcomes (see lemma 1 of the appendix). In general he/she will respond with a probability vector $p \in M(0)$. We are interested in scoring rules which reward the expert for responding truthfully, that is which reward the expert for responding with $p = q^{(N)}$. We shall first investigate scoring rules having this property for all N, and subsequently study rules which have this property in some asymptotic sense. E_Q denotes expectation with respect to Q. We assume $Q \in M(X)$.

DEFINITION. A *scoring rule for average probabilities* is a real valued function $R: M(0) \times N \times \Omega \to \mathbf{R}$.

NOTATIONAL CONVENTION. When $R(p, N, \omega)$ depends on ω only through the sample distribution $s^{(N)}$, we shall write $R(p, N, s^{(N)})$.

As we shall be exclusively concerned with scoring rules of this form, we drop the superscript (N), and simply write $R(p, N, s)$.

DEFINITION. For M a subset of $M(X)$, $R(p, N, s)$ is *positive* (*negative*) *sensed and M-strictly proper* if for all $Q \in M$

argmax (argmin) $E_Q R(p, N, s)$ is unique and equals $q^{(N)}$.
$p \in M(0)$.

The argmax (argmin) is taken over all non-degenerate probability vectors over the outcome set 0. $R(p, N, s)$ is called strictly proper if it is $M(X)$-strictly proper. Strict propriety is stronger for scoring rules for average probabilities than for scoring rules for individual variables, as the set $M(X)$ from which the assessor's probability is taken is larger than the set $M(0)$ from which the 'response distribution' is drawn.

THEOREM 1: *With the notation as above, let* $R(p, N, s)$ *be differentiable in p. Then the following are equivalent*:

(3)   *for all* $Q \in M(X)$, $\nabla_p E_Q(R(p, N, s))|_{p = q^{(N)}} = 0$.

(4)   *for i, j, k* $\in \{1, \ldots m - 1\}$, *there exist integrable functions* g_i, *and* g_{ikj}, *such that*:

$$(\partial/\partial p_i) R(p, N, s) = g_i(p, N)(p_i - s_i) +$$

$$+ \sum_{k < j} g_{ikj}(p, N)(s_k p_j - p_k s_j).$$

Remark 1. If R is differentiable in p, then (3) is a necessary condition for strict propriety. If R strictly convex or strictly concave in p, then (3) is also sufficient for strict propriety (with the rule's sense chosen appropriately).

Remark 2. Scoring rules of the sort described in this section are applicable to the quantile tests for calibration. In such tests the expert is asked to state quantiles corresponding to the probabilities $0 = f_0 < f_1 < f_2 <, \ldots f_{M-1} < f_M = 1$; of continuously distributed variables $X'_1, \ldots X'_N$. The variable X_i is then said to take outcome j if $X'_{i,j-1} < X'_i \leqslant X'_{i,j}$, where $X'_{i,j}$ denotes the j-th quantile of X'_i. If the expert is asked to state his probabilities for these outcomes then a strictly proper rule for average probabilities encourages him to respond with the probabilities $f_j - f_{j-1}, j = 1, \ldots M$.

Remark 3. If the outcome set is $\{0, 1\}$, then the variables $X_1, \ldots X_N$ are indicator functions for uncertain events. In this case the terms $(s_k p_j - p_k s_j)$, $k \neq j$ vanish and (4) takes the form:

(4') $(\partial/\partial p_i)R(p, N, s) = g_i(p, N)(p_i - s_i).$

Remark 4. Under the conditions of Remark 3, it follows from Theorem 1 that if R satisfies (4'), then:

$$R(p, N, s) = s \int_0^p g(x)\,\mathrm{d}x - \int_0^p xg(x)\,\mathrm{d}x.$$

A 'dual' equation appears in Savage (1971):

$$G(p, x) = -p \int_x^1 f(y)\,\mathrm{d}y + \int_x^1 yf(y)\,\mathrm{d}y$$

$G(p, x)$ is interpreted as the income of a subject who states price x when his true price is p, for commodities of which the experimenter will buy $f(y)$ units at price y, $0 \leqslant y \leqslant 1$.

EXAMPLES

The following proposition shows that the relative information score

$$R(p, N, s) = I(s^{(N)}, p) = \sum_{i=1}^m s_i^{(N)} \ln(s_i^{(N)}/p_i)$$

is strictly proper. The second statement shows that the roles of s and p cannot be reversed in this respect.

PROPOSITION 1. *Let* $s, p \in M(0)$;

(i) $I(s, p)$ *is a convex function of* p; *and* $p_m = 1 - \sum\limits_{i=1}^{m-1} p_i$

$$(\partial/\partial p_i)I(s, p) = \frac{p_i - s_i + \Sigma_{j=1,\ldots m-1}(s_i p_j - s_j p_i)}{p_i(1 - p_1 - \ldots p_{m-1})};$$

$i = 1, \ldots m - 1.$

(ii) *for* $m = 2$, *writing* $I(s, p)$ *as a function of* s_1, p_1, *and dropping the subscripts*:

$$(\partial/\partial s)I(s, p) = \ln((s - sp)/(p - sp))$$

$$= (s - p)/(p - p^2) + o(s - p); \quad (s - p) \to 0.$$

where $f(y) = o(y)$ *means that* $f(y)/y \to 0$ *as* y *approaches some limit* (*in this case* 0).

The last statement in Proposition 1 shows that $I(p, s)$ is *not* a strictly proper scoring rule.

A natural choice for R is:

$$R(p, N, s) = \sum_{i=1}^{m-1} c_i(s_i - p_i)^2,$$

where the c_i are constants (letting i run from 1 to m yields an expression of the same form). From Remark 1 to Theorem 1 it is easy to verify that this is a strictly proper scoring rule. Moreover, it corresponds to a 'quadratic loss function', where R is the loss incurred when one takes 'action' p while s is the 'true value'. Indeed, scoring rules for average probabilities can be regarded as loss functions for the random variable s. The set of negatively sensed proper scoring rules for average probabilities may be regarded as the set of loss functions for the random variable s which are minimized for the action p equal to the expected value of s.

The following three propositions extend the formalism of Theorem 1 to cover the case of an arbitrary finite number B of 'probability bins'. Σ_b denotes summation over the bins $b = 1, \ldots B$. We revive the notation of the previous section:

$p_{b,.}$ probability vector associated with bin b; $p_{b,.} \in M(0)$
$s_{b,.}$ sample distribution associated with bin b
$\mathbf{p} = p_{1,.}, \ldots p_{B,.}$
$\mathbf{s} = s_{1,.}, \ldots s_{B,.}$
n_b number of variables assigned to bin b
$\mathbf{n} = n_1, \ldots n_B$; $\mathbf{n}$ is called the occupation vector
$N = \Sigma_b n_b$.

The assessor performs his assessment by assigning each X_i, $i = 1$, $\ldots$; to one of the B bins. The vector of sample distributions s will be understood to depend on **n**, but we suppress this dependence in the notation and write $R(\mathbf{p}, \mathbf{n}, \mathbf{s}) := \Sigma_b R_b(p_{b,\cdot}, n_b, s_{b,\cdot})$.

PROPOSITION 2. *If* $R(\mathbf{p}, \mathbf{n}, \mathbf{s}) := \Sigma_b R_b(p_{b,\cdot}, n_b, s_{b,\cdot})$ *for scoring rules* R_b *for average probabilities having the same sense, then R is strictly proper if and only if* R_b *is strictly proper,* $b = 1, \ldots B$.

Proof. This follows immediately from the additivity of expectation:

$$E(R) = \Sigma_b E(R_b).$$ ∎

In particular, this proposition applies if the R_b's are all the same scoring rule. However, taking $B = 1$, then in general:

$$R(\mathbf{p}, \mathbf{n}, \mathbf{s}) \neq \sum_{i=1}^{N} R(p, 1, x_i).$$

This can readily be verified for the relative information score, taking $N = 2$, $M = \{0, 1\}$, $x_1 = 1$, $x_2 = 0$, $p = 1/2$. The right hand side is $2 \ln 2$, whereas the left hand side is 0. This emphasizes the difference between using scoring rules for average probabilities, as against using scoring rules for individual variables and adding the scores.

PROPOSITION 3. *If* $R(\mathbf{p}, \mathbf{n}, \mathbf{s})$ *is a strictly proper scoring rule, then so is*

$$R^{\sim}(\mathbf{p}, \mathbf{n}, \mathbf{s}) = R(\mathbf{p}, \mathbf{n}, \mathbf{s}) + f(\mathbf{s}, \mathbf{n}),$$

where f is an arbitrary real function on the sample and occupation vectors.

Proof. Assume that R is positively sensed. Since $f(\mathbf{s}, \mathbf{n})$ does not depend on **p**, taking expectation with respect to $Q \in M(X)$

$$\sup_{\mathbf{p}} ER^{\sim}(\mathbf{p}, \mathbf{n}, \mathbf{s}) = \sup_{\mathbf{p}} ER(\mathbf{p}, \mathbf{n}, \mathbf{s}) + Ef(\mathbf{s}, \mathbf{n})$$

$$= ER(\mathbf{q}, \mathbf{n}, \mathbf{s}) + Ef(\mathbf{s}, \mathbf{n}) = ER^{\sim}(\mathbf{q}, \mathbf{n}, \mathbf{s}). \blacksquare$$

PROPOSITION 4. *Let* $f(y, z)$ *be any real valued function on* $\mathbf{R}^2$ *with non-zero derivatives in y and z, and let R be any differentiable strictly proper scoring rule for average probabilities. Let* $H = H(\mathbf{n}, \mathbf{p})$ *be a real valued function satisfying* $\nabla_p H \neq 0$, $b = 1 \ldots B$. *Then* $f(R, H)$ *is not strictly proper.*

3. ASYMPTOTIC PROPERTIES

In this section we revert to the formalism of Theorem 1, involving one probability bin and we study asymptotic properties of scoring rules for average probabilities. A strong and weak form of asymptotic propriety are distinguished. Results with the weak form are easier to prove, and seem sufficient for applications. Results using the stronger form to date are more restricted. As before, M denotes a subset of $M(X)$. The definitions will be formulated for positively sensed rules, for negatively sensed rules, 'argmin' replaces 'argmax'.

DEFINITION. $R(p, N, s)$ is *strongly asymptotic M-strictly proper* if for all $Q \in M$,

$$\operatorname*{argmax}_{p} E_Q R(p, N, s) = p_N \text{ for some (not necessarily unique) } p_N,$$

$$\text{and if } q^{(N)} \to r \text{ as } N \to \infty, \text{ then also } p_N \to r \text{ as } N \to \infty.$$

$R(p, N, s)$ is *weakly asymptotic M-strictly proper* if for all $Q \in M$, whenever

$$q^{(N)} \to r \quad \text{as } N \to \infty,$$

and

$$r' \in M(0), \quad r' \neq r,$$

then there exists $N' \in \mathbf{N}$ such that for all $N > N'$;

$$E_Q R(r, N, s) > E_Q R(r', N, s).$$

The difference between strong and weak asymptotic propriety may be characterized as follows. For strong asymptotic propriety we first take the argmax over possible assessments, and take the limit as $N \to \infty$ of these argmax's. In weak asymptotic propriety we first choose an assessment r'. If r' is not equal to the limit of the average probabilities then it is possible to choose an N' in $\mathbf{N}$ such that for all $N > N'$, the expected score under r' is worse than the expected score using the limits of the average probabilities.

We now fix M to be the set of non-degenerate product probabilities for X; that is, $P \in M$ implies $P = p^\infty$ for some $p \in M(0)$. Under P the variables $X_1, \ldots$; become independent and identically distributed with distribution p. Our goal in this section is to show that a simple Fisher-type test of significance, and a weighted test of significance are weakly asymptotic M-strictly proper scoring rules for average probabilities.

Before turning to this, we first examine the asymptotic distributions under P, for $P \in M$, of the strictly proper scoring rules for average probabilities introduced in the previous section. We define:

$$C(p, N, s) = 2NI(s, p).$$

Statisticians recognize C as the log likelihood ratio; it has an asymptotic x^2 distribution with $m - 1$ degrees of freedom under P. C is also a strictly proper scoring rule for average probabilities, for all $P \in M(X)$. Moreover, if we have B probability bins, $B > 1$, then we can simply add the scores C for each bin. The resulting sum will have an asymptotic Chi-square distribution with $B(m - 1)$ degrees of freedom (we must also assume the variables in different bins are independent under P).

If we expand the logarithm in C and retain only the dominant term, we arrive at

$$\sum_{i=1}^{m} N(p_i - s_i)^2 / p_i.$$

which is the familiar Chi-square variable for testing goodness of fit. This has the same asymptotic properties as C, but is not a strictly proper scoring rule for average probabilities. The terms p_i in the denominators cause the gradient in (3) to have a term in $(p_i - s_i)^2$.

For $B = 1$, $m = 2$, the quadratic score has tractable properties. Put $p = p_1$, $s = s_1$. Under $P = p^\infty$ the variable

$$QU(p, N, s) = (p - s)^2$$

approaches a squared normal variable with mean 0 and variance $(p - p^2)/N$. QU is also strictly proper. If QU is standardized by dividing by the variance, the result will no longer be strictly proper. Without standardization, the distribution of sums of scores QU for $B > 1$ will not be tractable.

If we desire a test statistic which is also a strictly proper scoring rule for average probabilities, we are well advised to confine attention to C.

We define a simple and a weighted test of significance using C. $x^2_{(m-1)}$ denotes the cumulative Chi-square distribution function with $m-1$ degrees of freedom.

$$w_t(p, N, s) = \begin{bmatrix} 1 & \text{if} & C(p, N, s) \leqslant t \\ 0 & \text{if} & C(p, N, s) > t \end{bmatrix}$$

$$W_t(p, N, s) = [1 - x^2_{(m-1)}(C(p, N, s))] w_t(p, N, s).$$

The score w_t reflects simple significance testing. As we know the (asymptotic) distribution of $C(p, N, s)$ under suitable hypotheses, we can choose t such that the probability that $C(p, N, s) > t$ is less than some fixed number α. t is then called the *critical value* and α the *significance level* for testing the hypothesis that the n observations are independently drawn from the distribution p. W_t distinguishes 'degrees of acceptance' according to the probability under P of seeing a relative information score in N observations at least as large as $C(p, N, s)$.

PROPOSITION 5. *For $t \in (0, \infty)$, the score w_t is weakly asymptotic M-strictly proper.*

PROPOSITION 6. *If $t \in (0, \infty)$, then W_t is weakly asymptotically M-strictly proper.*

Propositions 5 and 6 show that simple and graduated significance testing constitute asymptotically strictly proper reward structures under suitable conditions. Weighing expert probability assessors with these scores is similar to treating the experts as classical statistical hypotheses. However, weighing expert assessments differs in two important respects from testing hypotheses. First, good expert assessments must not only 'pass statistical tests', but must have high information, or equivalently, low entropy. Second, the choice of significance level, or equivalently, the choice of t, is not determined by the same considerations which apply in hypothesis testing. We return to this point in the last section.

PROPOSITION 7. *For all $t \in (0, \infty)$, and for any function $f: M(0) \times N \to [a, b]$ with $0 < a < b < \infty$, the scores*

$$w_t(p, N, s)f(p, N); \qquad W_t(p, N, s)f(p, N)$$

are weakly asymptotic M-strictly proper.

Remark. The definition of f in this proposition is quite arbitrary, so long as it does not depend on Ω and so long as its range is bounded and bounded away from zero. Referring to Remark 2 of Theorem 1, f might depend on the distributions of the variables X'_{ii} instead of $M(0)$.

The proofs of Propositions 5 and 6 nowhere use the propriety of the score C. Had we used any other 'goodness of fit' statistic to define w and W, these proofs would still go through. For strong asymptotic propriety the case seems to be different; one result involving strong asymptotic propriety is stated in (Cooke, 1990).

4. A MENU OF WEIGHTS

In this section we consider weights for combining probability distributions, based on the theory of scoring rules. This section will use the notation of Propositions 2−4. That is, we consider the case where the variables X_i take values in the set of outcomes $0 = \{o_1, \ldots o_m\}$, and the subject assigns each variable to one of B 'probability bins' with assessed distribution $p_{b,} \in M(0)$, $b = 1, \ldots B$. As noted in Remark 2 to Theorem 1, quantile tests can be constitute a special case with $B = 1$.

DEFINITION. For a given finite set of indicator variables, a *weight for an expert assessment p* is a non-negative, positively sensed scoring rule for the average probabilities. A *system of weights for a finite set of experts* (perhaps assessing different indicator variables) is a normalized set of weights for each expert, if one of the experts' weight is positive; otherwise the system of weights is identically zero.

The above definition explicitly accounts for the eventuality that all experts might receive zero weight. Based on the discussions of the previous sections, we formulate four desiderata for a system of weights. Such weights should
  (1) reward low entropy and good calibration
  (2) be relevant
  (3) be asymptotically strictly proper (under suitable assumptions)
  (4) be meaningful, prior to normalization, independent of the specific variables from which they are derived.

The last desideratum is somewhat vague, but is understood to entail that the unnormalized weights for experts assessing different variables can be meaningfully compared.

The requirement of asymptotic strict propriety requires explanation.[1] Let us assume that an expert experiences 'influence on the beliefs of the decision maker' as a form of reward. Let us represent the decision maker's beliefs as a distribution P_{dm}, which can be expressed as some function G of experts' weights w_e and assessments P_e, $e = 1, \ldots E$:

$$P_{dm} = G(w_1, \ldots w_E; P_1, \ldots P_E).$$

Expert e's influence is found by taking the partial derivative of P_{dm} with respect to that argument which e can control, namely P_e. Maximizing expected influence in this sense is not always the same as maximizing the expected value of w_e. However, if G is a weighted average:

$$P_{dm} = \sum_{e=1}^{E} w_e P_e / K; \quad K = \sum_{e=1}^{E} w_e;$$

then

$$\partial P_{dm} / \partial P_e = w_e / K.$$

When giving his assessment, expert e will not generally know the weights of the other experts; he may not even know who they are or how many they are. Therefore the normalization constant K is effectively independent of the variable P_e which e can manipulate. Maximizing the expected influence $\partial P_{dm} / \partial P_e$ is effectively equivalent to maximizing the (unnormalized) weight w_e. Arguments for and against the above form for G are reviewed in French (1985) and Genest and Zidek (1986). In Cooke (1990) it is argued that weighted averaging is the only serious candidate for combining expert probability assessments.

Requirement 2 is satisfied by scoring rules for average probabilities in the following sense: the scores do not depend on the probabilities of outcomes which might have been observed but were not. Of course, the average of probabilities is not itself the probability of an outcome which can be observed. However, if the average of the probabilities of independent events converges to a limit, then the observed relative frequencies converge with probability one to the same limit.[2]

The scores w_t and W_t introduced in the previous section may be considered as weights which incorporate the notion of significance testing. They reward good calibration, but they are not sensitive to entropy.

Before discussing weights which are sensitive to entropy, we must distinguish the two notions of entropy.

(5) $H(\mathbf{n}, \mathbf{s}) = (1/N) \sum_b n_b H(s_{b,.})$: 'sample entropy'

(6) $H(\mathbf{n}, \mathbf{p}) = (1/N) \sum_b n_b H(p_{b,.})$: 'response entropy'.

where

$$H(p_{b,.}) = - \sum_{i=1}^{m} p_{b,i} \ln(p_{b,i})$$

and similarly for $H(s_{b,.})$. In conjunction with Remark 2 to Theorem 1, we note that the distinction between sample and response entropy is not meaningful for quantile tests, as $B = 1$ in this case. We introduce and discuss several weights using the notation of Proposition 2.

Proposition 3 allows us to add an arbitrary function of the sample distribution to a proper scoring rule. The maximal value of the sample entropy $H(\mathbf{n}, \mathbf{s})$ is $\ln(m)$. We can define positive, positively sensed weights taking values in $[0, 1]$ as follows:

(7) $w_t + \ln(m) - H(\mathbf{n}, \mathbf{s})$

(8) $W_t + \ln(m) - H(\mathbf{n}, \mathbf{s})$.

These weights bear a resemblance to the De Groot-Fienberg decomposition of the logarithmic scoring rule. Two criticisms may be directed against them. First, these weights can be zero only if w_t, respectively W_t are zero *and* if $H(\mathbf{n}, \mathbf{s})$ assumes the value $\ln(m)$. Hence, a poorly calibrated expert can still receive substantial weight. Moreover, the maximal value $\ln(m)$ can only be obtained if exactly half of the test events occur. Second, the term $H(\mathbf{n}, \mathbf{s})$ is also a random variable.

Different numerical values for $H(\mathbf{n}, \mathbf{s})$ cannot be meaningfully compared without taking the distribution of $H(\mathbf{n}, \mathbf{s})$ into account.

These objections can at least be partially met. Instead of considering the sample entropy $H(s_{b_.})$ for each bin, we could consider the relative information $I(s_{b_.}, S)$, where S denotes the total sample distribution over all variables. If the variables are independent and distributed according to S, the quantity

$$D(n, s) = \sum_b 2n_b I(s_{b_.}, S)$$

approaches x_B^2 in distribution as n goes to infinity. Large values of $D(n, s)$ indicate a 'high resolution of the base rate' which would be unlikely to result from statistical fluctuations under the distribution S. $x_B^2(D)$ will be termed the *base rate resolution index*. The following thus represent improvements over the weights (7) and (8):

(9) $w_t + x_B^2(D)$

(10) $W_t + x_B^2(D)$.

The weights (7)–(10) are easily seen to be asymptotically strictly proper in the sense of Proposition 7. For large n, $1 - x_B^2(D)$ repreesnts the significance level at which we would reject the hypothesis that the expert had assigned test variables to bins randomly. These weights still have the property that poorly calibrated experts can receive substantial weight.

Weights using a multiplicative entropy penalty based on the response entropy can avoid this problem. A suitable form for such weights is

(11) $w_t / H(\mathbf{n}, \mathbf{p})$

(12) $W_t / H(\mathbf{n}, \mathbf{p})$.

If the experts' weights are derived from assessment of *different* variables in the past, then comparing their response entropies might not be meaningful. For example, suppose one expert has assessed the probability of rain in the Netherlands, where the weather is quite unpredictable, and another has assessed the probability of rain in Saudi Arabia. The latter would have a lower response entropy simply because his assessments would all be near zero. If these experts' assessments were to be combined via weights derived from their past assessments, then their responses entropies should not be used. In such situations the term $1/H(\mathbf{n}, \mathbf{p})$ should be replaced by:

$$(13) \quad (1/N) \sum_b n_b I(p_b, S),$$

where S is the overall sample distribution (considered non-stochastic). (13) is the average information in the expert's assessments relative to the base-rate. To satisfy the conditions of Proposition 7, (13) must be bounded, and bounded away from zero.

These weights are zero whenever the expert's calibration score exceeds the critical value. Moreover, it is possible to argue that the response entropy is a more appropriate index for lack of information than the sample entropy. The weighted combinations refer indeed to the assessed probabilities $p_{b,.}$, and not to the sample probabilities $s_{b,.}$. If $\mathbf{s} = \mathbf{p}$, then these two coincide. The weak asymptotic propriety of these weights is proved in proposition 7. In the case of quantile tests, the term $1/H(\mathbf{n}, \mathbf{p})$ would simply be a constant and should be replaced in (11) and (12) by some other appropriately bounded function of the assessment.

5. HEURISTICS OF WEIGHING

We conclude with some heuristic remarks on choosing a score for weighing expert opinion. First, the unnormalized weights for each expert depend only on the assessments of each expert, and the realizations. When a decision maker calls various experts together for advice on a particular variable, he could compute weights based on prior assessments of *different* variables. When experts are combined, their weights must be normalized to add to unity. However, the decision maker must ensure that the experts are all calibrated on the same effective number of variables. In the language of hypothesis testing, this ensures that the experts are subjected to significance tests of equal power.

It seems obvious that we do not want to assign high weight to experts who are very poorly calibrated, regardless how low their entropy is, and regardless which information measure is used. Entropy should be used to distinguish between experts who are more or less equally well calibrated. The weight (12) does in fact behave in this way. The calibration scores W_t will typically range over several orders of magnitude, while the entropy scores typically remain within a factor 3. Because of the form of the weight (12), when weights of different experts are normalized, the calibration term will dominate, unless all experts are more or less equally calibrated. The weight (10) does not have this property. We also note that there are other measures of 'lack of information' which can be substituted for the function f in Proposition 7. Any other measure should be chosen in such a way that the calibration term will dominate in (12).

In testing hypotheses, it is usual to say that an hypothesis is 'rejected' when the test statistic exceeds its critical value. This way of speaking, of course, is not appropriate in dealing with expert probability assessments. In hypothesis testing, choice of the significance level α means choosing to reject true hypotheses with frequency α. In combining expert opinions, we are not trying to find the 'true' expert and 'reject' all of the others as 'false'. Indeed, if we collect enough data, then we will surely succeed in rejecting all experts, as no one will be *perfectly* calibrated. The significance level α can be chosen so as to optimize the decision maker's calibration and information scores.

Five full scale applications of (12), or suitable adaptations, in the aerospace and process sectors are described in (Cooke, 1990). In addition, probabilistic weather forecasts over the period 1980—1986 have been retroactively analysed with this model. In all cases the 'optimized decision maker' outperforms the best expert, and also outperforms the result of simple arithmetic averaging, where performance is measured in terms of the weight (12) which, of course, is also a scoring variable.

APPENDIX

This appendix gives the proofs for various results cited in the text.

THEOREM 1. *With the notation as above, let* $R(p, N, s)$ *be differentiable in p. Then the following are equivalent*:

(14)   *for all* $Q \in M(X), \nabla_p E_Q(R(p, N, s))|_{p = q^{(N)}} = 0.$

for i, j, k $\in \{1, \ldots m - 1\}$, *there exist integrable functions* g_i, *and* g_{ikj}, *such that*:

(15) $(\partial/\partial p_i)R(p, N, s) = g_i(p, N)(p_i - s_i) +$

$$+ \sum_{k<j} g_{ikj}(p, N)(s_k p_j - p_k s_j);$$

The proof uses three lemmata.

LEMMA 1. *Let $A_1, \ldots A_N \in F$ with indicator functions $1_1, \ldots 1_N$. Then*

$$(1/N) \sum_{i=1}^{N} Q(1_j = 1)$$

$$= (1/N) \sum_{k=1}^{N} kQ(\text{exactly } k \text{ of } A_1 \ldots A_N \text{ occur}).$$

Proof. The right hand side is just the expected relative frequency of $1_1, \ldots 1_N$, and is equal to $E((1/N) \sum 1_i) = (1/N) \sum E(1_i)$, which is equal to the left hand side. ∎

LEMMA 2. *Let A be a $L \times n$ matrix with rank L, $L < n$, and let $b \in \mathbf{R}^L, b \neq 0$. Let*

$$X = \{x \in \mathbf{R}^n | Ax = b, x_i > 0, \quad i = 1, \ldots n\}.$$

Suppose $X \neq \phi$, then the subspace generated by X has dimension $n - L + 1$.

Proof. For this proof, let $|X|$ denote the dimension of the subspace generated by X. Let Y be the subspace of vectors orthogonal to each $x \in X$. It suffices to show that $|Y| = L - 1$. Let $Z = \{z \in \mathbf{R}^n | Az = 0\}$. Then $|Z| = n - L$. Pick $x^{(o)} \in X$. We can find a basis $e^{(1)}, \ldots e^{(L)}$ of Z such that $x_i^{(o)} + e_i^{(l)} > 0$, $i = 1, \ldots n$, and thus $x^{(o)} + e^{(l)} \in X$, $l = 1, \ldots L$. For all $y \in Y$:

$$0 = \sum x_i^{(o)} y_i + \sum e_i^{(l)} y_i;$$

or

$$\sum x_i^{(o)} y_i = -\sum e_i^{(l)} y_i; \quad l = 1, \ldots L.$$

The left hand side does not depend on l, or on the choice of basis, hence both sides must vanish, and y must be orthogonal to $x^{(o)}$ and to all $z \in Z$. Since $b \neq 0$, $x^{(o)} \notin Z$, and y is orthogonal to $n - L + 1$ linearly independent vectors. The proof is completed by noting that any solution to the linear system $Ax = b$ may be written $x = x^{(o)} + z$ for some $z \in Z$. Hence $y \in Y$ if and only if y is orthogonal to $x^{(o)}$ and to $e^{(l)}, l = 1 \ldots L$, and $|Y| = L - 1$.

LEMMA 3. *Let $R(p, N, s)$ satisfy (14), and fix N. Let*

$$W = \left\{ k \in \mathbf{N}^{m-1} | k_i \geq 0, i = 1, \ldots m - 1, \sum_{i=1}^{m-1} k_i \leq N \right\}.$$

For $Q \in M(X)$, let

$$Q(k_{i_1}, \ldots k_{i_L}) = Q\{ \text{outcome } i_j \text{ occurs } k_{i_j} \text{ times in} \\ x_1, \ldots x_N | j = 1, \ldots L \}$$

Use $k \in W$ to index the coordinates of $\mathbf{R}^{|W|}$. Then writing

$$Q_k = Q(k); \quad k \in W$$

we may consider each element of $M(X)$ as a vector in $\mathbf{R}^{|W|}$. Without loss of generality, we write $R(p, N, s) = R(p, N, k/N)$, where $k \in W$. For each differentiable scoring rule, for each $p \in M(0)$ and each $i = 1, \ldots m - 1$ we may consider $R(p, i)$ with coordinates

$$R(p, i)_k = (\partial/\partial p_i)R(p, N, k/N); \quad k \in W,$$

as an element of $\mathbf{R}^{|W|}$. For $p \in M(0)$ let

$$A(p) = \text{the subspace of } \mathbf{R}^{|W|} \text{ generated by} \\ \{ Q \in M(X) | q^{(N)} = p \}.$$

In other words, $A(p)$ is the subspace generated by probability vectors Q whose vector of average probabilities equals p. Let

$$B(p, i) = \text{the subspace of } \mathbf{R}^{|W|} \text{ generated by the vectors} \\ R(p, i) \text{ where the scoring rule } R \text{ is differentiable in} \\ p \text{ and satisfies (15).}$$

then for all $p \in M(0)$, and $i = 1, \ldots m - 1$;

$$\text{Dim } A(p)^\perp \leq \text{Dim } B(p, i),$$

where $A(p)^\perp$ denotes the subspace of vectors in $\mathbf{R}^{|W|}$ orthogonal to $A(p)$.

Proof. We first show that Dim $A(p)^\perp = m - 1$. Letting Σ' denote summation over W, the vectors $Q \in A(p)$ are in the positive cone and satisfy the m linear equations:

$$\Sigma' \, Q(k) = 1;$$

$$\Sigma' \, k_i Q(k) = Np_i; \quad i = 1, \ldots m - 1.$$

The following consideration shows that these equations are independent: Since for $Q \in M(X)$, $q_i^{(N)} > 0$, $i = 1, \ldots m$; we can always find two probability vectors Q, Q' whose vectors of average probabilities $q^{(N)}$ and $q'^{(N)}$ disagree in just one coordinate j, $1 \leqslant j \leqslant m - 1$. By Lemma 2, the dimension of $A(p)$ equals $|W| - (m - 1)$, hence the dimension of $A(p)^\perp$ equals $m - 1$.

We show that Dim $B(p, i) \geqslant m - 1$. Fix i. If $m = 2$, Dim $A(p)^\perp = 1$, the functions g_{ij} are all zero, and it is trivial to show that $B(p, i)$ has dimension 1. Assume $m \geqslant 3$. It suffices to find $m - 1$ linearly independent vectors in $B(p, i)$. In fact, it suffices to find $m - 1$ vectors whose components on $m - 1$ coordinates $k^{(1)} \ldots k^{(m-1)}$ are linearly independent, where:

$$k_h^{(j)} = N\delta_{jh}; \quad \delta_{jh} = 1 \text{ if } j = h \text{ and } = 0 \text{ otherwise.}$$

It suffices to find scoring rules $R^{(1)} \ldots R^{(m-1)}$ satisfying (15) such that the $(m - 1)$ by $(m - 1)$ matrix Y with

$$y_{jh} = (\partial/\partial p_i)R^{(j)}(p, N, k^{(h)}/N))$$

has full rank. We choose

$$R^{(i)}(p, N, s) = 1/2p_i^2 - p_i s_i$$

$$R^{(j)}(p, N, s) = -s_j \ln p_j - s_i \ln p_i$$

$$- (1 - s_i - s_j)\ln(1 - p_i - p_j); \quad j \neq i.$$

It is easy to verify that

$$\partial/\partial p_i R^{(i)} = p_i - s_i;$$

$$\partial/\partial p_i R^{(j)} = (p_i - s_i + s_i p_j - p_i s_j)/p_i(1 - p_i - p_j)$$

$$\partial/\partial p_j R^{(j)} = (p_j - s_j + s_i p_j - p_i s_j)/p_j(1 - p_i - p_j),$$

other derivatives being zero. This shows that the scores $R^{(j)}$ satisfy (15); $j = 1 \ldots m - 1$. Multiplying row j by $p_i(1 - p_i - p_j)$ for $j \neq i$. The matrix Y has the form

$$
\begin{array}{cccccccc}
k^{(1)} & \cdots\cdots\cdots\cdots\cdots\cdots\cdots\cdots\cdots\cdots & k^{(i)} & \cdots\cdots\cdots & k^{(m-1)} \\
\\
0 & p_i & p_i & p_i \ldots p_1 + p_i - 1 & \ldots\ldots & p_i & \\
p_i & 0 & p_i & p_i \ldots p_2 + p_i - 1 & \ldots\ldots & p_i & \\
p_i & p_i & 0 & p_i \ldots p_3 + p_i - 1 & \ldots\ldots & p_i & \\
\cdot \\
\cdot \\
\cdot \\
p_i & p_i & p_i & p_i \ldots p_i - 1 & \ldots\ldots & p_i & \text{i-th row} \\
\cdot \\
\cdot \\
p_i & p_i & p_i & p_i \ldots p_{m-1} + p_i - 1 & \ldots\ldots & 0.
\end{array}
$$

That Y has full rank can be seen by subtracting the i-th row of the above matrix from each of the other rows. The result is

$$
\begin{array}{cccccccc}
-p_i & 0 & 0 & 0 \ldots p_1 & 0 & 0 \ldots 0 \\
0 & -p_i & 0 & 0 \ldots p_2 & 0 & 0 \ldots 0 \\
0 & 0 & -p_i & 0 \ldots p_3 & 0 & 0 \ldots 0 \\
p_i & p_i & p_i & p_i \ldots p_i - 1 & p_i & p_i \ldots p_i \\
0 & 0 & 0 & 0 \ldots p_{i+1} & -p_i & 0 \ldots 0 \\
0 & 0 & 0 & 0 \ldots p_{m-1} & 0 & 0 \ldots -p_i
\end{array}
$$

These rows are linearly dependent if and only if

$$ p_1 + p_2 + \ldots + p_{i-1} + p_{i+1} + \ldots + p_{m-1} = 1 - p_i; $$

however, this condition cannot hold if $p_m > 0$, which is the case if $p \in M(0)$. It follows that Y has full rank, and the proof is completed. ∎

PROOF OF THEOREM 1. We fix N and adopt the notation of Lemma 3.

(15) implies (14):

(16) $(\partial/\partial p_i)E_Q R(p, N, s) = \sum' Q(k)\,(\partial/\partial p_i)R(p, N, k/N)$

$$= \sum' Q(k)\,[g_i(p, N)\,(p_i - (k_i/N)) +$$

$$+ \sum_{f<j} g_{ifj}(p, N)\,((k_f/N)p_j - p_f(k_j/N))]$$

$$= g_i(p, N) \sum_{k_i=0}^{N} Q(k_i)\,(p_i - (k_i/N)) +$$

$$+ \sum_{f<j} g_{ifj}(p, N) \sum_{k_f, k_j = 0}^{N} Q(k_f, k_j) \times$$

$$\times [(k_f/N)p_j - p_f(k_j/N)].$$

If $p = q^{(N)}$, the last expression equals 0 by Lemma 1.

(14) implies (15):

Let R be a differentiable scoring rule satisfying (14). Then for all $Q \in M(X)$

$$(\partial/\partial q_i(E_Q R(q, N, s) = \sum' Q(k)\,(\partial/\partial q_i)R(q, N, k/N) = 0.$$

It follows that for all $q \in M(0)$ and $i = 1, \ldots m - 1$, $R(q, i) \in A(q)^{\perp}$. From Equation (16) it follows that $B(q, i)$ is contained in $A(q)^{\perp}$. From Lemma 3 it now follows that $B(q, i) = A(q)^{\perp}$, hence $R(q, i) \in B(q, i)$. Since this holds for all $q \in M(0)$, $i - 1, \ldots m - 1$, and since R is differentiable, it follows that R has the form (15). ∎

PROOF OF PROPOSITION 1

Proof. The statements regarding the partial derivatives can be verified by direct calculation. To verify that $I(s, p)$ is convex, it suffices to verify for $r \in (0, 1)$, $p_1, p_2 \in M(0)$; $p = rp_1 + (1 - r)p_2$, that $rI(s, p_1) + (1 - r)I(s, p_2) \geqslant I(s, p)$. We have:

$$I(s, p) = \sum_{i=1}^{m} s_i \ln(s_i) - \sum_{i=1}^{m} s_i \ln p_i.$$

The first term is the negative entropy of s, and is always non-positive. It suffices to verify that

$$\sum_{i=1}^{m} s_i(r \ln p_{1,i} + (1-r) \ln p_{2,i}) \leq \sum_{i=1}^{m} s_i \ln p_i,$$

which indeed follows immediately from the concavity of the function $\ln(x)$.

To verify the estimate in the last equation of (ii), write $s = p + \theta p$. Then:

$$s/p = 1 + \theta; \quad \theta = (s - p)/p;$$
$$(1 - s)/(1 - p) = 1 - \theta p/(1 - p).$$

Using the Taylor expansion, valid for $x \in (-1, 1)$,

$$\ln(1 + x) = x - x^2/2 + x^3/3 \ldots;$$
$$\ln(s/p) - \ln((1 - s)/(1 - p)) = \theta + \theta p/(1 - p) + o(\theta)$$
$$= (s - p)/(p - p^2) + o(\theta). \quad \blacksquare$$

PROOF OF PROPOSITION 4

Proof. Fix b and put $p_{j,.} = q_{j,.}$ for $j \neq b$. Then R and f may be considered as scoring rules for the b-th bin, with R strictly proper. Without loss of generality we therefore put $B = 1$, drop the subscript b, and show that in this case $f(R, H)$ cannot be proper. Let '∂_i' denote partial derivation with respect to p_i, $i = 1 \ldots m - 1$. If $f(R, H)$ were strictly proper we should have by Theorem 1:

$$\partial_i f(R, H) = (\partial f/\partial R)\partial_i R + (\partial f/\partial H)\partial_i H$$

$$= g_i(p, n)(p_i - s_i) + \sum_{k < j} g_{ikj}(p, n)(s_k p_j - s_j p_k);$$

for suitable functions g_i, g_{ikj}. Since R is differentiable and strictly proper,

$$\partial_i R = h_i(p, n)(p_i - s_i) + \sum_{k < j} h_{ikj}(p, n)(s_k p_j - s_j p_k);$$

for suitable functions h_i, h_{ikj}. Hence:

$$(\partial f/\partial H)\partial_i H = (p_i - s_i)(g_i - h_i \, \partial f/\partial R) +$$

$$+ \sum_{k < j} (s_k p_j - s_j p_k)(g_{ikj} - h_{ikj} \, \partial f/\partial R).$$

Since $\partial_i H$ and $(\partial f/\partial H)$ do not depend on s, this can hold only if both sides are identically zero. However, the left hand side cannot equal zero for all values of p, since $(\partial f/\partial H) \neq 0$, $\partial_i H(p, n) \neq 0$. Therefore $f(R, H)$ cannot be strictly proper. ∎

PROOF OF PROPOSITION 5

Proof. Choose $Q \in M$. Then $\lim_{N \to \infty} E_Q w_t(q, N, s) = x^2_{m-1}(t) > 0$. Choose $r \in M(0)$ with $r \neq q$. By the strong law of large numbers, $s \to q$ Q-almost surely, and by Egoroff's theorem, for every $d > 0$, the convergence is uniform on a set of Q-probability greater than $1 - d$. Choose $d < x^2_{m-1}(t)$. For some $I_d > 0$ we can find $N_d \in \mathbf{N}$ such that on this set, for all $N > N_d$, $I(s, r) > I_d$. For $N > \max\{N_d, t/I_d\}$, on this set

$$2NI(s, r) > t.$$

Hence for sufficiently large N, $Q\{w_t(r, N, s) = 0\} > 1 - d$ and:

$$E_Q w_t(r, N, s) < d < x^2_{m-1}(t). \quad ∎$$

This argument also shows that $E_Q w_t(r, N, s) \to 0$ as $N \to \infty$. The proof of Proposition 6 uses the following lemma.

LEMMA 4. *For any (right continuous) cumulative distribution function F and any $z \in \mathbf{R}$,*

$$0 \leq \int_{-\infty}^{z} F(x)\,dF(x) - F(z)^2/2 \leq \max_x(F(x) - F_-(x));$$

where $F_-(x) = \sup_{y < x} F(y)$.

Proof. Let 1_A denote the indicator function of the set A. Since F is bounded, F is integrable with respect to dF and the Fubini theorem may be applied.

$$\int_{-\infty}^{z} F(x)\, dF(x) = \iint 1_{(-\infty, z]}(x)\, 1_{(-\infty, x]}(y)\, dF(y)\, dF(x)$$

$$= \iint 1_{[y, z]}(x)\, 1_{(-\infty, z]}(y)\, dF(x)\, dF(y)$$

$$= \int (F(z) - F(y) + F(y) -$$

$$- F_-(y))\, 1_{(-\infty, z]}(y)\, dF(y)$$

$$= F(z)^2 - \int_{-\infty}^{z} F(y)\, dF(y) +$$

$$+ \int_{-\infty}^{z} (F(y) - F_-(y))\, dF(y).$$

The two inequalities now follow from the fact that

$$0 \leqslant \int (F(y) - F_-(y))\, dF(y) \leqslant \max_x (F(x) - F_-(x)). \qquad \blacksquare$$

COROLLARY

$$\int_{-\infty}^{z} F(x)\, dF(x) \geqslant F(Z)^2/2$$

with equality holding for all z if and only if F is continuous.

PROOF OF PROPOSITION 6: Choose $Q \in M$. Let Q_N denote the expert's cumulative distribution function for $C(q, N, s)$.

$$E_Q W_t(q, N, s) = \int_0^t (1 - x_{m-1}^2(x))\, dQ_N$$

$$= E_Q w_t(q, N, s) - \int_0^t x_{m-1}^2(x)\, dQ_N.$$

$Q_N \to x_{m-1}^2$. x_{m-1}^2 is continuous and bounded, so the Helly Bray theorem in conjunction with the corollary to Lemma 4 yield, as $N \to \infty$:

$$\int_0^t x_{m-1}^2(x)\, dQ_N(x) \to \int_0^t x_{m-1}^2(x)\, dx_{m-1}^2(x)$$

$$= x_{m-1}^2(t)^2/2.$$

Hence, $\lim_{N \to \infty} E_Q W_t(q, N, s) = x_{m-1}^2(t) - x_{m-1}^2(t)^2/2 > 0$. Suppose $r \in M(0)$ with $r \neq q$. Since $1 - x_{m-1}^2(t) < 1$, it follows from the proof of Proposition 5 that

$$\lim_{N \to \infty} E_Q W_t(r, N, s) = 0. \qquad \blacksquare$$

Remark. Note that the continuity of x_{m-1}^2 is essential in the above proof. If we replaced x_{m-1}^2 in the definition of W_t by a non-continuous distribution, for example Q_N, then Proposition 6 would yield only a crude overestimate of $E_Q W_t$. This illustrates the advantage of studying propriety from the asymptotic perspective. Note also that $t \neq \infty$ is essential.

PROOF OF PROPOSITION 7. We treat only the first score, as the argument for the second is similar. Choose $Q \in M$. Suppose $r \in M(0)$, $r \neq q$. We must show that for all sufficiently large N;

$$E_Q w_t(q, N, s)/E_Q w_t(r, N, s) > f(r, N)/f(q, N).$$

The right hand side is bounded from above by b/a, and this bound does not depend on N. From the proof of Proposition 5, the left hand side goes to ∞ as $N \to \infty$. $\qquad \blacksquare$

Delft University of Technology, The Netherlands

NOTES

[1] This discussion is inspired by comments in Genest and McConway (1989).

[2] Since according to Kolmogorov's sufficient condition for the strong law of large

numbers (Tucker, 1967, p. 124), for indicator functions 1_i,

$$p(\lim_{n \to \infty} (1/n) \sum_{i=1}^{n} (1_i - E(1_i)) = 0) = 1.$$

Hence, if the average probabilities $(1/n) \sum E(1_i)$ converge to p as $n \to \infty$, then with probability one

$$\lim_{n \to \infty} (1/n) \sum_{i=1}^{n} 1_i = p.$$

BIBLIOGRAPHY

Alpert, M. and Raiffa, H. 1982. 'A Progress Report on the Training of Probability Assessor'. In D. Kahneman, P. Slovic, and A. Tversky (eds.), *Judgment under Uncertainty: Heuristics and Biases*, Cambridge University Press, pp. 294—306.

Apostolaris, G. 1988. 'Expert Judgment in Probabilistic Safety Assessment'. In C. Clarotti and D. Lindley (eds.), *Proceedings International School of Physics "Enrico Fermi"*, Amsterdam.

Bohla, B., Blauw, H., Cooke, R., and Kok, M. 1988. 'Expert Resolution in Project Management'. Paper presented at the Conference on Decision Making under Risk, Budapest, June 5—10.

Clarotti, C. and Lindley, D. (eds.). 1988. Accelerated Life Testing and Experts' Opinions in Reliability, *Proceedings of the International School of Physics "Enrico Fermi"*. Amsterdam.

Cooke, R. 1985. 'Expert Resolution'. *Proc. of the IFAC Conference*. Varsese Italy, September 1985, Permagon Press.

Cooke, R., Mendel, M., and Thys, W. 1988. 'Calibration and Information in Expert Resolution: A Classical Approach'. *Automatica* **24** 87—94.

Cooke, R. 1988. 'Uncertainty in Risk Assessment'. *Reliability Engineering and System Safety* **23** 277—285.

Cooke, R. 1989. 'A Theory of Weights for Combining Expert Opinion'. Report to the Dutch Ministery of Environment. *Expert Opinion in Safety Studies*, vol. iv, Delft.

Cooke, R., Mendel, M., and Van Steen, J. 1989. 'Model Description Report, Report to the Dutch Ministery of Environment. *Expert Opinion in Safety Studies*, vol. iii. Delft.

Cooke, R. 1990. *Experts in Uncertainty*. Appearing Oxford University Press.

De Finetti, B. 1937. 'La prevision: ses lois logiques, ses sources subjectives'. *Annales de l'Instut Henri Poincare* **7** 1—68. English translation in Kyburg and Smolker (eds.), *Studies in Subjective Probability*. Wiley.

De Groot, M. 1974. 'Reaching Consensus', *J. Amer. Statis. Assoc.*, vol. 69, 118—121.

De Groot, M. and Fienberg, S. 1986. 'Comparing Probability Forecasters: Basic Binary Concepts and Multivariate Extensions'. In P. Goel and A. Zellner (eds.), *Bayesian Inference and Decision Techniques*, Elsevier.

De Groot M. and Fienberg, S. 1983. 'The Comparison and Evaluation of Forecasters'. *The Statistician* **32** 12—22.

De Groot, M. and Bayarri, M. 1987. 'Gaining Weight: A Bayesian Approach'. Dept. of Stat. Carnegie Mellon University Tech. Report 388, Jan. 1987.

Feller, W. 1971. *An Introduction to Probability Theory II*, New York: Wiley.

French, S. 1985. 'Group Consensus Probability Distributions', In J. Bernado, M. De Groot, D. Lindley, and A. Smith (eds.), *Bayesian Statistics*, Elsevier, pp. 188—201.

Friedman, D. 1983. 'Effective Scoring Rules for Probabilistic Forecasts'. *Management Science* **29** 447—454.

Genest, C. and McConway, K. 1989. 'Allocating the Weights in the Linear Opinion Pool'. Quebec: Universitè Laval.

Genest, C. and Zidek, J. 1986. 'Combining Probability Distributions: A Critique and Annotated Bibliography'. *Statistical Science* **1** 114—148.

Honglin, W. and Duo, D. 1985. 'Reliability Calculation of Prestressing Cable System of the PCPV Model'. *Trans. of the 8th Intern. Conf. on Structural Mech. in Reactor Techn*. Brussels, Aug. 19—23, pp. 41—44.

Hoel, P. 1971. *Introduction to Mathematical Statistics*. New York: Wiley.

Lichtenstein, S., Fischoff, B., and Phillips, D. 1982. 'Calibration of Probabilities: The State of the Art of 1980'. In D. Kahneman, P. Slovic, and A. Tversky (eds.), *Judgment under Uncertainty: Heuristics and Biases*, Cambridge University Press, pp. 306—335.

Matheson, J. and Winkler, R. 1976. 'Scoring Rules for Continuous Probability Distributions'. *Management Science* **22** 1087—1096.

McCarthy, J. 1956. 'Measures of the Value of Information'. *Proc. of the National Academy of Sciences*, pp. 654—655.

Morris, P. 1977. 'Combining Expert Judgments: A Bayesian Approach', *Management Science*, vol. 23 nr. 7, 679—693.

Murphy, A. 1973. 'A New Vector Partition of the Probability Score'. *J. of Applied Meteorology* **12** 595—600.

Preyssl, C. and Cooke, R. 1989. 'Expert Judgment; Subjective and Objective Data for Risk Assessment of Spaceflight Systems', *Proceedings PSA '89 International Topical Meeting Probability, Reliability and Safety Assessment*, Pittsburgh, April 2—7.

Roberts, H. 1965. 'Probabilistic Prediction'. *J. Amer. Statist. Assoc.* **60** 50—62.

Savage, L. 1971. 'Elicitation of Personal Probabilities and Expectations'. *J. Amer. Statis. Assoc.* **66** 783—801.

Shuford, E., Albert, A., and Massengil, H. 1966. 'Admissible Probability Measurement Procedures'. *Psychometrika* **31** 125—145.

Stael von Holstein, C. 1970. 'Measurement of Subjective Probability'. *Acta Psychologica* **34** 146—159.

Tucker, H. 1967. *A Graduate Course in Probability*. New York: Wiley.

Wagner, C. and Lehrer, K. 1981. *Rational Consensus in Science and Society*, Dordrecht: Kluwer Acad. Publ.

Winkler, R. and Murphy, A. 1968. 'Good Probability Assessors'. *J. of Applied Meteorology* **7** 751—758.

Winkler, R. 1968. 'The Consensus of Subjective Probability Distributions'. *Management Science* **15** B61—B75.

Winkler, R. 1969. 'Scoring Rules and the Evaluation of Probability Assessors'. *J. Amer. Statist. Assoc.* **64** 1073—1078.

Winkler, R. 1986. 'On Good Probability Appraisers'. In P. Goel and A. Zellner (eds.), *Bayesian Inference and Decision Techniques*, Elsevier.

ETTORE MARUBINI

SHORT AND LONG TERM SURVIVAL ANALYSIS IN ONCOLOGICAL RESEARCH

1. INTRODUCTION

Chronic diseases and cancer among these are, nowadays, the main concern of medical research in developed countries. Chronic disease has several characteristics which lead to its complexity and make it difficult to trace its 'natural history' and to design clinical trials aiming at evaluating the effectivness of therapeutical strategies. Among these characteristics there are a multifactorial aetiology, nearly non specific and often exogenous, and a long time course with continuous interactions between the host biological system and the surrounding environment. As a consequence, clinical researches accomplished to study cancer treatment usually last a considerable time period and sometimes, as for instance in breast cancer, several years. Their duration may be partitioned into a recruitment stage in which different treatments are administered and a stage of continued observation after accrual has stopped. During both these stages a great amount of information concerning variables possibly capable of influencing the duration, course and outcome of the disease (i.e. prognostic factors) has to be collected *ad hoc* or has to be attained from data gathered as a subsidiary aspect of an ongoing study. Due allowance for these factors must be made in the statistical analysis performed to evaluate the treatments effectivness in order to enlighten their prognostic role and to make the analysis more sensitive (Armitage and Gehan, 1974; Marubini *et al.*, 1983; Harrel *et al.*, 1985).

In such studies the response variables most widely used are: length of survival and/or length of disease-free survival and nearly always they come with a feature which creates special problems in the analysis of the data. This feature is known as censoring and, broadly speaking, it occurs when there are individuals in the sample for whom only a lower (or upper) bound on lifetime is available.

The statistical methodology of survival data has seen notable development in past fifteen years or so. Historically, results were often given in terms of the percentage of patients surviving some arbitrary time period, usually five years; this clearly underutilizes the information

R. Cooke and D. Costantini (eds), Statistics in Science. The Foundations of Statistical Methods in Biology, Physics and Economics, 73—87.

provided by the data. The introduction of life-table (Berkson and Gage, 1950) and product limit (Kaplan and Meier, 1958) methods enable more flexible data description, and formal methods of statistical inference, for example Gehan's (1965) and logrank (Mantel, 1966) tests, followed. A further aspect was the statistical modelling of survivorship, aiming at a more sophisticated processing of data than can be achieved by simple survival curves description and significance testing. Attention focuses here upon given prognostic variables and the model, either parametric or quasi-parametric (see Lawless, 1982; Aranda-Ordaz, 1983; Tibshirani and Ciampi, 1983) serves as a mean for evaluating the relative effects of each of them, their interrelationship and their combined action.

In diseases with a relatively low fatality rate, for example thyroid and breast cancer, one more issue of extreme concern is the assessment of '*curability*' in long-term survival studies. The statistical definition of this term, which is not the same as the clinical concept of curability, leads to consider a set of patients '*cured*' if, after a given number of years from the diagnosis, they assume the age-specific death rates of the general population. This implies to resort to methods of *relative survival analysis* which make allowance of the expected deaths in a group of people in the general population who are similar, for sex and age, to the patient group at the beinning of the follow-up period.

Section 2 of this paper aims at unifying, in the same frame, the models most widely used in clinical literature to process both survival data and relative survival data and briefly commenting upon their main features. In Section 3 the main characteristics of a breast cancer case series followed-up for more than 20 years are given and in Section 4 results obtained by analysing survivorship of this series are shown. Section 5 contains discussion and concluding remarks.

2. REGRESSION MODELS

Consider failure time $T > 0$ and suppose that a vector $\mathbf{x}' = (x_1, x_2, \ldots, x_i, \ldots, x_p)$ of concomitant variables has been observed for each of the n subjects. Note that $\mathbf{x}$ may include both quantitative and qualitative variables such as treatment group; this latter may be inserted through the use of dummy variables. The problem consists in modelling and investigating the relationship between t and $\mathbf{x}$.

2.1. *Survival Analysis*

A first class of models is defined in terms of the effect produced by the covariates on the underlying hazard function. Namely:

(1)   *multiplicative model*: $h_j(t/\mathbf{x}) = h(t)\, g(\mathbf{x}_j)$

(2)   *additive model*: $h_j(t/\mathbf{x}) = h(t) + g'(\mathbf{x}_j)$

of explanatory variables, of the j-th patient; $h(t)$ is the underlying hazard when $g(\mathbf{x}) = 1$ or $g'(x) = 0$ depending on the model, and $g\,(\,\cdot\,)$ and $g'\,(\,\cdot\,)$ are parametric functions of the covariates. Aranda-Ordaz (1983) discussed a family of transformations which includes models (1) and (2) as special cases and suggested to use such a family to define a scale suitable for investigating the consistency with the data of the two models.

A particularly useful set of models is obtained from model (1) by defining $g(\mathbf{x}) = \exp(\mathbf{x}'\,\boldsymbol{\beta})$ where: $\mathbf{x}'\boldsymbol{\beta} = x_{1j}\beta_1 + x_{2j}\beta_2 + \ldots + x_{ij}\beta_i + \ldots + x_{pj}\beta_p (i = 1, 2, \ldots p)$ and the β_i's are unknown coefficients. This form is usual in regression analysis and sufficiently flexible for many purposes. Thus model (1) can be rewritten in its form widely used in clinical literature:

(3) $h_j(t/\mathbf{x}) = h(t)\exp(\mathbf{x}_j'\,\boldsymbol{\beta})$.

If one assumes a particular form for $h(t)$, a fully parametric proportional hazard model is obtained; the most frequently used among such models is the Weibull one for which: $h(t) = \theta\sigma(\theta t)^{\sigma-1}$ where $\theta > 0$ and $\sigma > 0$ are parameters of scale and shape respectively. Weibull models include the exponential model as the special case $\sigma = 1$. The parameters of the hazard function and the vector $\boldsymbol{\beta}$ are estimated by resorting to the maximum likelihood method.

Cox (1972) suggested to leave $h(t)$ unspecified and to regard it as a nuisance function for the purpose of inference on the βs. On somewhat heuristic grounds he suggested a '*partial likelihood*' function $L(\boldsymbol{\beta})$ for estimating $\boldsymbol{\beta}$ in (3) in the absence of knowledge of $h(t)$. The concept of partial likelihood was presented by Cox (1975) who showed that the full likelihood function can be factored in two terms:

(4) $L[\boldsymbol{\beta}, h(t)] = L_1(\boldsymbol{\beta})\, L_2[\boldsymbol{\beta}, h(t)]$.

The first of them, being independent of $h(t)$, allows the user to estimate

β and, by making the additional assumption that $h(t)$ is constant between successive failures, to estimate it by means of $\hat{\beta}$ (Oakes, 1972; Breslow, 1974).

A deep discussion of the partial likelihood approach and of the loss of information involved in using it is out of the scope of this paper; the reader is referred to papers by Cox (1975), Efron (1977), Aalen (1978) and Oakes (1981).

Methods proposed for assessing the assumption underlying the Cox's model have been reviewed and commented upon by Kay (1984).

A second important class of models (accelerated failure time models, AFT) assumes that the vector **x** acts linearly on $Y = \ln T$ and hence multiplicatively on T. Suppose that Y is related to **x** via a linear model:

$$(5) \qquad Y_j = \mathbf{x}_j' \beta + \varepsilon_j$$

where ε_j is an error variable with a given density. Exponentiation gives $T = T' \exp(\mathbf{x}_j' \beta)$ where $T' = \exp(\varepsilon) > 0$ has hazard function $h(t')$ which is independent of β. The hazard function for T can be written in terms of this base-line hazard $h(t')$ (see: Marubini and Valsecchi, 1987):

$$h_j(t/x) = h\left[t \exp(-\mathbf{x}_j' \beta)\right] \exp(-\mathbf{x}_j' \beta).$$

AFT models can be built up by assuming that ε has normal, logistic or generalized gamma density functions. As for the parametric models of the previous class the maximum likelihood method is adopted to estimate the parameters of the density functions and the vector β.

2.2. *Relative Survival Analysis*

In any long-term follow-up study of a sample of cancer patients the survival pattern reflects mortality not only from the disease under study but also from deaths due to all other causes. The risk of death from causes other than that under study varies with sex, age and with calendar time. The relative survival analysis adjusts for the expected mortality and thus it makes possible meaningful comparisons of the survival experience of groups of patients which differ for sex, age and calendar period of observation.

Models till now suggested for handling relative survival can be obtained by inserting a time dependent function $[\lambda(t)]$ in model (1). Namely:

(6) $\quad h_j(t/\mathbf{x}) = h(t)\, g(\mathbf{x}_j) + \lambda_j(t)$

(7) $\quad h_j(t/\mathbf{x}) = h(t)\, g(\mathbf{x}_j) \cdot \lambda_j(t)$

where, for $t = 0$, $\lambda_j(0)$ is the mortality rate expected in the general population for people who are similar, for sex and age to the j-th patient at the beginning of the follow-up interval. Thus, in (6) $h(t) \cdot g(\mathbf{x}_j)$ denotes the excess mortality risk which depends on the force of mortality specific to the disease under study and on the explanatory variables. By assuming that the excess mortality risk decays exponentially with time, and that $g(\mathbf{x}_j) = \exp(\mathbf{x}_j'\, \boldsymbol{\beta})$, Pocock *et al.* (1982) suggested the following model for investigating the curability of breast cancer:

(8) $\quad h_j(t/\mathbf{x}) = \exp[\mathbf{x}_j'\, \boldsymbol{\beta} + \tau_1 - \tau_2(t - c)] + \lambda_j(t)$

where c is a given constant; $c = 5$ was chosen by Pocock *et al.* (1982) on the basis of the graphical analysis of the excess death rate in their series.

In the context of survival data grouped into a number, say K, of the subintervals $[t_k, t_{k+1}]$, $k = 1, 2, \ldots, K$, a similar model was adopted by Hakulinen and Tenkanen (1987) and by Hakulinen *et al.* (1987). Namely:

(9) $\quad h_{kj}(t/\mathbf{x}) = \exp\{\mathbf{x}_{ik}'\, \boldsymbol{\beta} + a_k(t)\} + \lambda_{kj}(t).$

The constant $a_k(t)$ depends on time $t \in [t_k, t_{k+1})$ and the values for the independent variables x_{ikj} may be different for different intervals $[t_k, t_{k+1})$, $k = 1, 2, \ldots, K$. It clearly appears that the force of mortality of the disease under study (i.e. the excess of mortality due to it): $\lambda_{kj}(t/\mathbf{x}) - \lambda_{kj}(t)$ follows a proportional hazard model (Cox, 1972). After assuming a binomial error function the constants a_k and $\boldsymbol{\beta}$ were estimated using the GLIM system.

With reference to time-aggregated survival data, Hill *et al.*, (1985) elaborated upon a model previously suggested by Breslow *et al.* (1983) to evaluate the carcinogenic effect of exposure to chemicals in cohort studies; it may be obtained from Equation (7) by postulating $h(t) = 1$

and $g(\mathbf{x}_j) = \exp(\mathbf{x}_j \boldsymbol{\beta})$:

(10) $h_{kj}(t/\mathbf{x}) = \exp(\mathbf{x}'_{kj}\boldsymbol{\beta}) \cdot \lambda_{kj}(t).$

In this way one models the ratio of the mortality rate in the patients sample to the mortality rate in the reference population.

An extension of model (10) which takes into account the exponential decay of $h(t)$ was suggested by Mezzanotte et al. (1987b) and an application of it will be shown in the next section. The model is:

(11) $h_{kj}(t/\mathbf{x}) = \exp(\mathbf{x}'_{kj}\boldsymbol{\beta} + \gamma t_k) \cdot \lambda_{kj}(t).$

From Equation (7) one can easely derive the semiparametric Cox-type (1972) model proposed by Andersen et al., (1985):

(12) $h_j(t/\mathbf{x}) = h(t)\exp(\mathbf{x}'_j\boldsymbol{\beta}) \cdot \lambda_j(t).$

Here $h(t)$ is an unknown underlying relative mortality, at time t, for a patient with $\mathbf{x}_j = \mathbf{0}$.

3. BREAST CANCER PATIENTS

At the Istituto Nazionale per lo Studio e la Cura dei Tumori in Milan, from January 1964 to January 1968, 737 patients with breast cancer were randomized to undergo either extended radical mastectomy (i.e. with internal mammary nodes dissection) or conventional Halsted mastectomy; 716 were considered evaluable. Patterns of relapse and survival of this series have already been published by Valagussa et al. (1987). Patients with non disseminated breast cancer classified as T_1, T_2, T_{3a}, N^- or N^+ (according to TNM classification, 1978) were eligible for the study. Patients in the two treatment groups were comparable in relation to the following prognostic factors: extent of the disease, menopausal status, location of primary, axillary involvement and age at surgery (Veronesi e Valagussa, 1981).

The data-base was updated at June 28th 1986 and the results of the following analyses refer to this date. After more than 20 years of follow-up, patients lost are only 4.75%. Since it was shown that the survival curves for the two surgical treatment groups were super-imposed each to the other, the statistical analyses were accomplished on the whole series of 716 patients.

4. RESULTS

4.1. *Survival Analysis*

Figure 1 gives the survival curve of the sample estimated according to the product limit method (Kaplan-Meier, 1958) and Figure 2 the relative hazard curve.

From a univariate preliminary analysis the prognostic role of menopausal status and of location of primary appeared to be negligible; therefore only metastases at the axillary nodes, extent of the disease and age at surgery were inserted into the regression model.

Cox's (1972), Weibull, log-logistic and log-normal models were fitted by the LIFEREG procedure (1985). The estimate of the regression coefficients and their standard errors are reported in Table 1 together with the code adopted to generate the regressors.

It may be easely seen that the values of the costants estimated by the four models are very similar, though, if one considers the three parametric models, the natural logarithm of the likelihood function of the log-logistic and lognormal models are very similar whilst they compare favorably to that of Weibull model. This finding is in agreement with the pattern of the estimated hazard function shown in Figure 2.

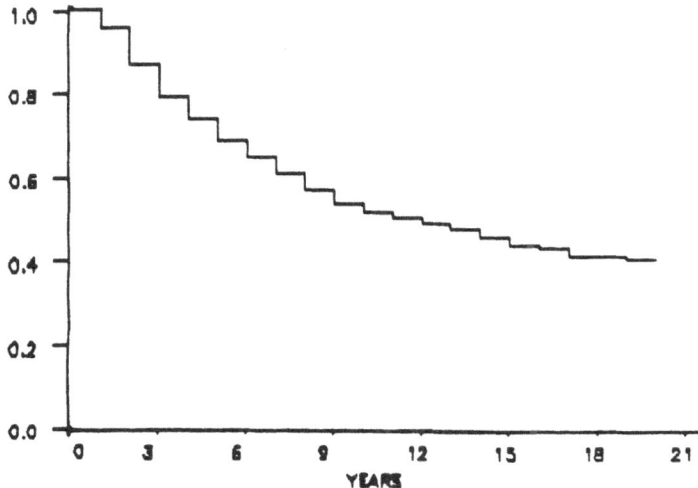

Fig. 1. Survival of the whole case series. Ordinate: cumulative probability.

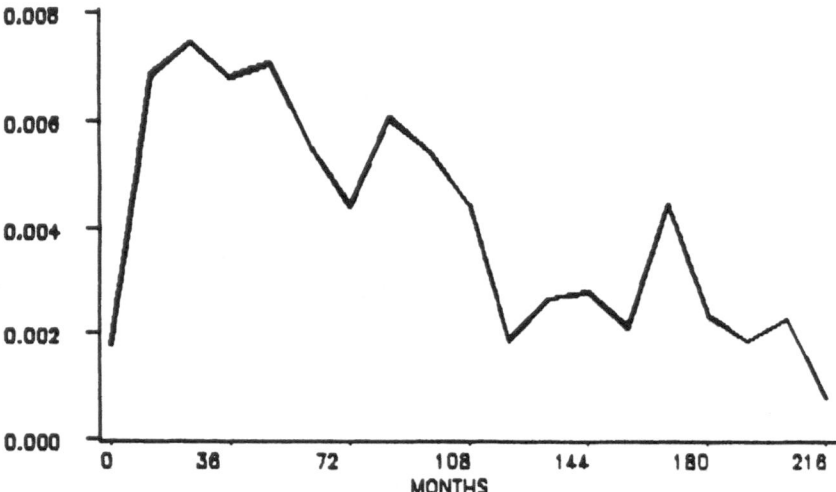

Fig. 2. Force of morality of the whole case series. Ordinate: instantaneous death probability.

TABLE 1

Survival analysis: regression coefficients obtained by means of different models

Variable and code	Cox's PH		Weibull		Log-logistic		Log-normal	
	β	s.e.*(β)	β	s.e.*(β)	β	s.e.*(β)	β	s.e.*(β)
Metastase, at the axillary notes								
$N-$ 0								
$N+$ 1	1.1477	0.1051	1.2117	0.1073	1.2736	0.1028	1.1978	0.1045
Extent of tumor								
T_1 0 1	−0.8310	0.2080	−0.8380	0.2059	−0.8550	0.2076	−0.9412	0.2126
T_2 1 0	−0.3728	0.1407	−0.3863	0.1388	−0.3151	0.1508	−0.3547	0.1540
T_3 0 0								
Age								
50—54 0								
Others 1	0.3515	0.1344	0.3908	0.1327	0.3532	0.1371	0.2873	0.1340
Intercept			−6.0034	0.1920	−5.5854	0.2022	−5.4644	0.2017
Scale			0.9877	0.0420	0.7225	0.0294	1.2443	0.0459
ln (partial) likelihood	−2589.36		−943.76		919.64		−915.10	

* s.e. (β): standard error of the estimate of the regression coefficient.

The goodness of fit of the three parametric models is investigated in Table 2 which gives the deaths predicted (F_k) by each of them for each year of follow-up together with the pertinent standardized residuals, $(z_k = (O_k - F_k)/\sqrt{F_k})$. It appears that the fitting of log-logistic and log-normal models is better that that of the Weibull one.

The score $-\mathbf{x}_j' \boldsymbol{\beta}$ computed on each patient with $\boldsymbol{\beta}$ vector estimated by means of the log-normal model enabled the researchers to partition the series in three subsets with clearly different survival patterns. The rationale and the technical details of the approach are given by Mezzanotte *et al.* (1987a); the survival curves of the three groups are

TABLE 2

Survival analysis. Goodness of fit of different models. O_x = observed deaths; F_x = deaths predicted by the relative model

Years after surgery	Observed deaths (O_x)	Model Weibull Predicted deaths (F_x)	z_x	Log-logistic Predicted death (F_x)	z_x	Log-normal Predicted death (F_x)	z_x
1	17	44.67	−4.14	33.48	−2.85	28.24	−2.20
2	54	44.08	1.49	47.68	0.92	49.97	0.57
3	58	39.97	2.85	46.83	1.63	47.86	1.47
4	47	34.98	2.03	41.56	0.84	41.47	0.86
5	39	31.18	1.40	36.63	0.39	36.00	0.50
6	32	27.68	0.82	31.82	0.03	31.01	0.18
7	24	25.18	−0.24	28.26	−0.80	27.42	−0.65
8	33	23.27	2.02	25.51	1.48	24.70	1.67
9	25	21.08	0.85	22.50	0.53	21.78	0.53
10	19	19.01	0.00	19.89	0.20	19.27	0.06
11	8	17.54	−2.28	17.93	−2.35	17.37	−2.25
12	12	16.74	−1.16	16.69	−1.15	16.17	−1.04
13	10	15.78	−1.46	15.36	−1.37	14.89	−1.27
14	8	14.88	−1.78	14.18	−1.64	13.75	−1.55
15	16	14.31	0.45	13.30	0.74	12.89	0.87
16	7	13.29	−1.73	12.07	−1.46	11.70	−1.37
17	7	12.48	−1.55	11.17	−1.25	10.82	−1.16
18	7	11.81	−1.40	10.33	−1.04	10.02	−0.95
19	2	9.75	−2.48	8.32	−2.19	8.08	−2.14
20	5	7.37	−0.87	6.17	−0.47	6.00	−0.41

reported in Figure 3. The curves estimated with the product-limit method (Kaplan and Meier, 1958) for each group are drawn together with the pertinent ones estimated by the log-normal model.

4.2. *Relative Survival Analysis*

In accord to the definition of '*curability*' mentioned in Section 1 and after noting that the estimated hazard reaches its maximum at about the third year of follow-up, the relative survival analysis was performed from this year onwards. Both the additive model (9) and the *ad hoc* developed multiplicative model (11) were fitted. As it was shown (Mezzanotte *et al.*, 1987b) that the fitting of model (11) compares favourably with that of model (9), attention will focus here on the results given by the former one. GLIM system was used to estimate the constants; a Poisson distribution of errors was specified, the natural link function was called upon and the $\ln[\lambda_{kj}(t)]$ was inserted in the model equation by means of the $ OFFEST command.

On the whole series the scaled deviance from the model = 17.96,

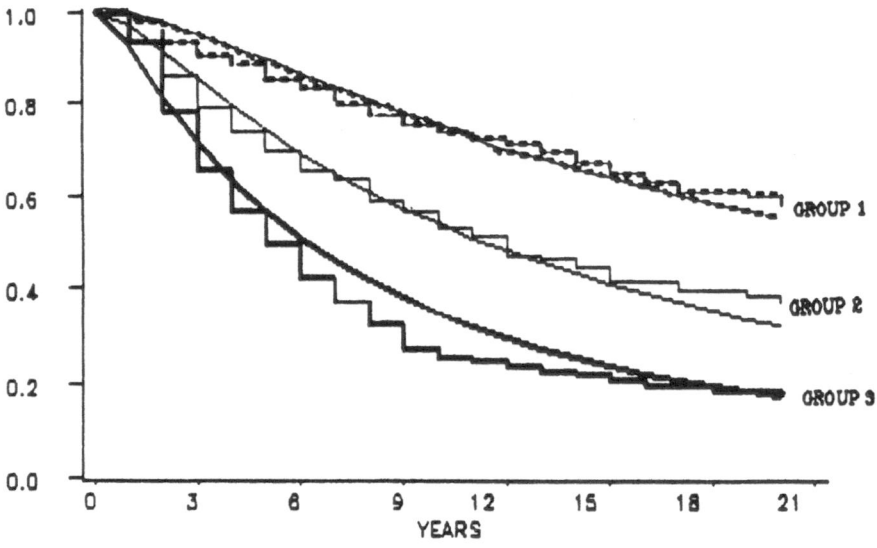

Fig. 3. Survival: product limit estimate *vs* log-normal model estimate. Ordinate: cumulative probability.

with 16 degrees of freedom (d.f.), is clearly not significant; Figure 4 depicts the pattern of relative survival.

After making allowance of the explanatory variables identifying the three groups with different prognosis elaborated upon in the previous subsection, the scaled deviance from the model is equal to 42.84 which, with 48 d.f. is still far from significance. In Figure 5 the relative survival curves predicted by the model (11) for each group are drawn.

Standardized residuals for each group are reported in Table 3 and show that Equation (11) models the observed data in a satisfactory way. In the same table the estimated index of '$curability$' = F_k/E_k is given. It is the ratio of the mortality fitted by the model to the mortality expected in a group of women in the general population who have the same age of the patient group at the beginning of the follow-up period. Since the model (11) is multiplicative, the expected value of this index for a '$cured$' person is one. Thus the results of the present analysis suggest that one can sensibly assess that a breast cancer patient is '$cured$' only after a minimum follow-up interval which ranges from 16 to 18 years according to the patient's characteristics observed at surgery.

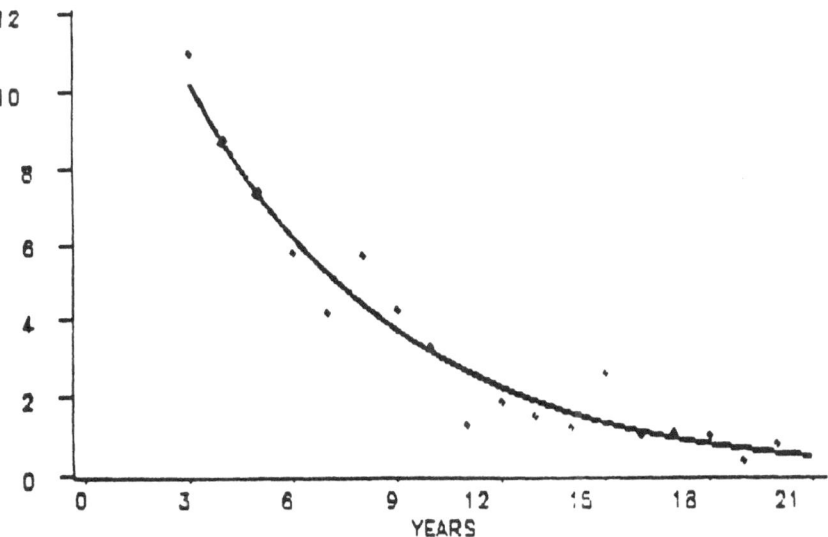

Fig. 4. Relative survival analysis of the whole series. On the ordinate: Index of Curability (see text).

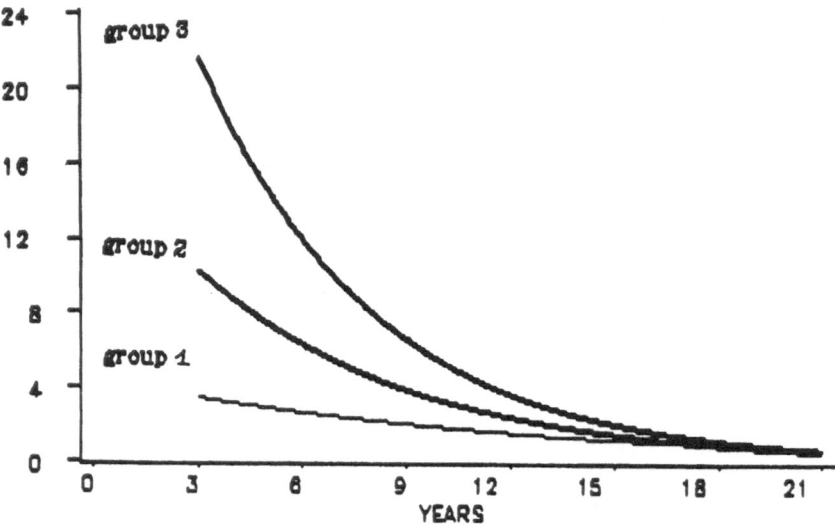

Fig. 5. Relative survival analysis of the three groups with different prognosis. Ordinate:
Index of Curability (see text).

5. DISCUSSION

There are two aspects to any of the models cited in Section 2 which deserve consideration: the form of $g(\mathbf{x})$ and the form of $h(t)$.

As in ordinary multiple regression analysis, trying to specify $g(\mathbf{x})$ implies firstly to consider which are the x_i variables to be entered into the equation and, secondly, which should their functional form be.

The body of evidence emerging from papers mentioned in the sub-setction 2.1. suggests that in significance testing little is to be gained by imposing restrictions upon $h(t)$.

As far as breast cancer patients survival is concerned, Pocock *et al.* (1982) and Gore *et al.* (1984) assess that Cox's, Weibull and log-logistic models give the same view of how the prognostic factors influence survival. Results shown in Table 1 agree with such an assesment.

On the other hand in cancer research, survival models are frequently used for predicting the time until some event and thus identifying groups of patients at a high risk of sustaining the event in a given time.

TABLE 3

Relative survival analysis. Goodness of fit after making allowance of the three groups. F_κ = deaths predicted by model (11); E_κ = deaths expected from mortality in the general population (see text)

t_κ	Group 1		Group 2		Group 3	
	Residual	F_κ/E_κ	Residual	F_κ/E_κ	Residual	F_κ/E_κ
3	0.3176	3.48	−0.5119	10.26	−0.0122	21.56
4	−0.8747	3.18	0.8920	8.78	−0.3376	17.68
5	1.0250	2.90	−2.2771	7.51	−0.4884	14.49
6	−0.0378	2.65	0.2138	6.43	−0.3882	11.88
7	−0.3394	2.42	−1.2770	5.50	−0.0244	9.75
8	1.2451	2.21	0.7344	4.70	1.3559	7.99
9	−0.4266	2.02	−0.5662	4.02	2.5548	6.55
10	0.0860	1.84	0.8343	3.44	−0.0218	5.37
11	1.3660	1.68	−1.1721	2.95	−0.7569	4.41
12	−0.8806	1.54	1.6129	2.52	−0.8187	3.61
13	−0.3928	1.40	−0.1264	2.16	−0.2055	2.96
14	−0.1507	1.28	0.0010	1.85	−1.2400	2.43
15	2.5219	1.17	1.1390	1.58	0.3266	1.99
16	−0.2539	1.07	−0.3179	1.35	−0.1152	1.63
17	−0.0960	0.98	−0.2793	1.16	0.2667	1.34
18	0.0237	0.89	−0.2435	0.99	0.4581	1.10
19	−1.6140	0.81	0.1791	0.85	0.1403	0.90
20	1.2420	0.74	−0.9036	0.72	0.9421	0.74

In such a context, when regression effects may be large or some physical aspects of the problem may be mirrored in a given form of $h(t)$, then it may be worth defining $h(t)$ parametrically and estimating the relative parameters. This approach was utilized in the Subsection 4.1. by resorting to the log-normal model which enabled the partition of the whole series in three subsets with an evident difference in their prognosis. This classification relies upon patients characteristics which do not reflect some peculiarity of the sample, but conform to the clinical evidence of the natural history of the disease. Anyway it seems sensible not to ascribe too rigid an interpretation to the estimated survival function in terms of long run frequencies since it will have been constrained either by the choice of $h(t)$ or of $g(x)$ or both (O'Quigley, 1982).

With regard to relative survival analysis, it appears from literature that the choice between the additive model (6) and the multiplicative model (7) rest mainly on the preference of the author(s) for one or other of the two strategies used by epidemiologists to evaluate the impact of a specific cause of death on a population in terms of '*excess mortality risk*' and of '*mortality rate ratio*' respectively.

Findings in the Subsection 4.2. are obtained by the multiplicative model (11); they appear slightly more '*optimistic*' than those published by Pocock *et al.* (1982) who found a significant excess mortality even 15 to 20 years after surgery, though the decreasing excess death rate in their series is consistent with an asymptotic null excess.

Università degli Studi di Milano, Milano, Italy

BIBLIOGRAPHY

Aalen, O. 1987. 'Nonparametric Inference for a Family of Counting Processes'. *Ann. Stat.* **6** 701—726.

Andersen, P. K., Borch-Johnsen, K., Deckert, T., Green, A., Hongaard, P., Keiding, N., and Kreiner, S. 1985. 'A Cox Regression Model for the Relative Mortality and its Application to Diabetes Mellitus Survival Data'. *Biometrics* **41** 921—932.

Aranda-Ordaz, F. J. 1983. 'An Extension of the Proportional-Hazards Model for Grouped Data'. *Biometrics* **39** 109—117.

Armitage, P. and Gehan, E. A. 1974. 'Statistical Methods for the Identification and Use of Prognostic Factors', *Cancer* **13** 16—36.

Berkson, J. and Gage, R. P. 1950. 'Calculation of Survival Rates for Cancer'. *Staff Meet. Mayo Clin* **25** 270—286.

Breslow N. E. 1974. 'Covariance Analysis of Censored Survival Data'. *Biometrics* **30** 89—99.

Breslow, N. E., Lubin, J. H., Marek, P., and Langholz, B. 1983. 'Multiplicative Models and Cohort Analysis. *J.A.S.A.* **78** 1—12.

Cox, D. R. 1972. 'Regression Models and Life Tables (with Discussion)'. *J.R.S.S.* B **34** 187—220.

Cox, D. R. 1975. 'Partial Likelihood. *Biometrika* **62** 269—276.

Efron B. 1977. 'The Efficiency of Cox's Likelihood Function for Censored Data. *J.A.S.A.* **72** 555—565.

Gehan, E. A. 1965. 'A Generalized Wilcoxon Test for Comparing Arbitrarily Singly-Censored Samples'. *Biometrika* **52** 203—223.

Gore, S., Pocock, S., and Kerr, G. R. 1984. 'Regression Models and Non-Proportional Hazards in the Analysis of Breast Cancer Survival'. *Appl. Statistics* **33** 176—195.

Hakulinen, T. and Tenkanen, L. 1987. 'Regression Analysis of Relative Survival Rates'. *Appl. Statistics* **36** 309—317.

Hakulinen, T., Tenkanen, L., Abeywickrama K., and Paivarinta L. 1987. 'Testing Equality of Relative Survival Patterns Based on Aggregated Data'. *Biometrics* **43** 313—325.

Harrel, F. E., Lee, K. L., Matchar, D. B., and Reichert, T. A. 1985. 'Regression Models for Prognostic Prediction: Advantages, Problems and Suggested Solutions'. *Canc. Treat. Rep.* **69** 1071—1077.

Hill, C., Laplanche, A., and Rezvani, A. 1985. 'Comparison of the Mortality of a Cohort with the Mortality of a Reference Population in a Prognostic Study'. *Stat. Med.* **4** 295—302.

Kalbfleisch, J. D. and Prentice, R. L. 1980. *The Statistical Analysis of Failure Time Data.* New York: J. Wiley and Sons.

Kaplan, E. L. and Meier, P. 1958. 'Nonparametric Estimation from Incomplete Observations'. *J.A.S.A.* **53** 457—481.

Kay, R. 1984. 'Goodness of Fit Methods for the Proportional Hazards Regression Model: A Review'. *Rev. Epidém. Santé Pub.* **32** 185—198.

Lawless, J. F. 1982. *Statistical Models and Methods for Lifetime Data.* New York: J. Wiley and Sons.

LIFEREG procedure (The) (1985). In: *SAS User's Guide: Statistics.* 507—528.

Mantel, N. 1966. 'Evaluation of Survival Data and Two New Rank Order Statistics Arising in its Consideration'. *Canc. Chemother. Rep.* **50** 163—170.

Marubini, E., Morabito, A., and Valsecchi, M. G. 1983. 'Prognostic Factors and Risk Groups: Some Results Given by Using an Algorithm Suitable for Censored Survival Data'. *Stat. Med.* **2** 295—303.

Marubini E. and Valsecchi M. G. 1987. *Analisi della sopravvivenza in sperimentazioni cliniche controllate e nelle osservazioni pianificate.* Monografia N 9 del Centro Zambon dell'Università degli studi di Milano.

Mezzanotte, G., Boracchi P., Valagussa, P., and Marubini, E. 1987a. 'Analisi della sopravvivenza a lungo termine nel carcinoma mammario'. *Rivista di Stat. Appl.* **20** 251—267.

Mezzanotte G., Boracchi, P., Marubini E., Valagussa P. and Veronesi, U. 1987b. 'Long Term Prognosis and Curability of Breast Cancer'. *Proceedings of the 8th International Meeting on Clinical Biostatistics,* Gotheborg, 7—11 September.

Oakes, D. 1972. 'Contribution to the Discussion of Paper by D. R. Cox.' *J.R.S.S.* B, **34** 202—205.

Oakes, D. 1981. 'Survival Times: Aspects of Partial Likelihood'. *Int. Stat. Rev.* **49** 235—264.

O'Quigley, J. 1982. 'Regression Models and Survival Prediction'. *The Statistician* **31** 107—116.

Pocock, S. J., Gore, S. M., and Kerr, G. R. 1982. 'Long Term Survival Analysis: the Curability of Breast Cancer'. *Stat. Med.* **1** 93—104.

Tibshirani, R. J. and Ciampi, A. 1983. 'A Family of Proportional and Additive-Hazards Models for Survival Data'. *Biometrics* **39** 141—147.

Union International Contre Le Cancer 1978. *TNM Classification of Tumors.* Geneva.

Valagussa, P., Bonadonna, G., and Veronesi, U. 1978. 'Patterns of Relapse and Survival Following Radical Mastectomy. Analysis of 716 Consecutive Patients'. *Cancer* **41** 1171—1178.

Veronesi, U. and Valagussa, P. 1981. 'Inefficacy of Internal Mammary Nodes Dissection in Breast Cancer Surgery'. *Cancer* **47** 170—175.

T. CALIŃSKI, E. OTTAVIANO AND M. SARI GORLA

A STATISTICAL APPROACH TO THE STUDY
OF POLLEN FITNESS

POLLEN FITNESS: IMPLICATIONS WITH REGARD TO POPULATION
STRUCTURE AND EVOLUTION

The theory of evolution based on natural selection assumes that individuals of a population differ with regard to adaptability to the environment where the population lives. As a consequence the contribution of each individual to the progeny generation, in terms of progeny size, differs. This contribution is referred to as fitness, and is expressed in a quantitative manner relative to a standard individual, generally the fittest of the population. Accordingly, fitness is a quantity varying from zero to one. The component characters producing this variability are: viability, reproductive ability and fertility.

When the variability in fitness is at least in part genetically controlled it produces evolution: genes carried by the fittest individuals increase in the population in each generation and, consequently, the adaptability of the population is improved. This concept has been stated by Fisher (1930) in his fundamental theorem of natural selection: 'the rate of increase in fitness of a population at any time is equal to its genetic variance in fitness at that time'.

However, the dynamics of population evolution of haploid and diploid species differ. In haploids, such as bacterial, since all the genetical differences are phenotypically expressed, the evolution rate can reach a maximum value. In diploid species, for example animals, genes having detrimental effects on fitness components are to a large extent found in heterozygous combinations: only a small proportion are in homozygous combinations and consequently the evolution rate is expected to be much lower then in haploids.

The main factor limiting a high evolution rate in both haploids and diploids is represented by the potential population size: a high evolution rate implies a high cost of selection (number of individuals eliminated by natural selection): when selection reduces the effective population size the process can lead to extinction.

Higher plant species are characterized by a life cycle consisting of a diploid phase and a haploid phase (the male and the female gameto-

R. Cooke and D. Costantini (eds), Statistics in Science. The Foundations of Statistical Methods in Biology, Physics and Economics, 89—101.
© 1990 *Kluwer Academic Publishers.*

phytes). Moreover, the male gametophytic generation attains a very large population size ($n \times 10^6$). For these reasons it has been proposed that if the variability in fitness is expressed in this generation, the population can attain a very high evolutionary rate without affecting the sporophytic population size.

To verify this hypothesis, which can explain the large increase of the Angiosperms in the vegetal world, two important biological assumptions have to be proved:

(i) that the male gametophytic population shows genetic variability for fitness;
(ii) that selection in this phase of the life cycle also affects the sporophytic population.

THE PROBLEM OF MEASUREMENT OF POLLEN FITNESS: RELATIVE POLLEN COMPETITIVE ABILITY

The measurement of pollen fitness and its variability is one of the most serious limitations in population genetics. The problem is even more difficult when dealing with gametophytic populations. One way to overcome some of the difficulties lies in the study of the variability and of the response to selection of component traits.

Male gametophytic fitness is to a large extent represented by pollen competitive ability, which is expressed when different pollen genotypes are in competition in the same style. The main components of the character are: pollen germination time, pollen tube growth rate and fertilization ability.

AN EXPERIMENTAL APPROACH

A possibility to measure and select for these components is given by plant species showing different style length. In fact, when different pollen grains are in competition in the same style, the probability of fertilization of the faster growing pollen tube increases with the length of the styles. If a mixture of two pollen genotypes is used, pollen competitive ability can be expressed as the proportion of the resulting progeny produced by each genotype.

The system also allows the application of gametophytic selection because the progeny produced by flowers having short styles and the progeny of long styled flowers are obtained at low and high gametophytic selection intensity, respectively.

In maize the inflorescence has a very favourable structure for this type of study. The same ear has hundreds of flowers and the style length changes according to the position of the flower on the ear, increasing from the top to the base. In this situation gametophytic competitive ability can be evaluated by a mixed pollination technique: pollen of a given line is mixed in equal proportions with the pollen of a standard line. If we can recognize the progeny produced by the two types of pollen, relative fitness can be expressed as proportion of progeny produced by the pollen of the line to be evaluated. Here the discrimination can be based on the colour of the kernels: the standard pollen produces coloured kernels.

In maize the special structure of the ear (the female inflorescence) allows the estimation of two parameters describing two different pollen fitness components:

(i)   relative pollen tube growth rate expressed by the regression of the proportions of one of the kernel types on ear segments;
(ii) viability and germination time as the proportion of one kernel type on the apex of the ear.

When this method is used to study families obtained under low and high selection intensity, data produced are as represented in Table 1.

From this set of data information can be obtained with regard to genetical variability of pollen tube growth rate (proportion of genetic variability accounting for differences between regressions fitted to each line), the proportion of genetic variability due to genes expressed in the haploid phase (Low vs. High). If plant characters are also studied, it is possible to measure the effect of MGS on the sporophytic generation. In other words, it is possible to test both biological assumptions.

STATISTICAL ANALYSIS

Frequencies of uncoloured kernels per segment of the ears obtained by

TABLE 1

Experimental data

Selection intensity	Replic.	Lines	Ear Segments					p-vector (600 × 1)
			1	2	3	4	5	

$$
\begin{array}{llll}
& & & \text{Ear Segments} & & & & & \\
\end{array}
$$

Low:
- Replic. 1, Line 1: p_{111}, p_{112}, $\cdots$, p_{115}
- Replic. 1, Line 2: p_{121}, p_{121}, p_{125}
- Replic. 2, Line 1: p_{211}, p_{221}, $\cdots$
- Replic. 2, Line 2: p_{221}, p_{222}

High:
- Replic. 1, Line 31: $p_{1\,31\,1}$, $\cdots$
- Replic. 1, Line 32: $p_{1\,32\,1}$
- Replic. 2, Line 31: $p_{2\,31\,1}$, $\cdots$
- Replic. 2, Line 32: $p_{2\,32\,1}$
- $p_{2\,60\,5}$

p-vector:

$$
\begin{bmatrix} p_{111} \\ \vdots \\ \\ = \quad p_{jik} \\ \\ \vdots \\ p_{2\,60\,5} \end{bmatrix}
$$

p_{jik}: proportions of fertilization of lines obtained by gametophytic selection (j denoting the replication, i the line and k the segment).

mixed pollination technique allow the estimation of two gametophytic parameters:

(i) the proportion of uncoloured kernels at the apex of the ear, which indicates the relative competitive ability of the line due to pollen viability and germinability,

(ii) the coefficient of linear regression of the proportion of uncoloured kernels on ear segments, which indicates the component of gametophytic competitive ability due to pollen tube growth rate.

A statistical procedure suitable for this information is based on regression analysis and variance component estimation. Data reported in the table are binomial proportions and the number of observations for each estimate is not constant. Therefore normality of distribution and homogeneity of error variances cannot be assumed.

Data transformation

With regard to the distribution distortion, the main interest being the linear component of the regression of proportions on ear segments, an appropriate transformation of the data was adopted. The vector $\mathbf{p}$ of observed proportions was linearly transformed, by means of an operator matrix $\mathbf{A}$, into a vector $\mathbf{F} = \mathbf{Ap}$, according to Grizzle, Starmer and Koch (1969). The $\mathbf{A}$ matrix, which specifies additive operations, was used to generate linear functions of the observed proportion vector $\mathbf{p}$. The matrix can be written as $\mathbf{A} = \mathbf{I} \otimes \mathbf{a}'$, where $\mathbf{I}$ is an identity matrix of order equal to the number of experimental units (or sub-populations), denotes Kronecker multiplication and $\mathbf{a}'$ is a row vector of the orthogonal polynomial coefficients which define the linear component; thus the analysis of the $\mathbf{F}$ variables refers only to the linear regression variability. Moreover, since the sample sizes for observing the proportions are sufficiently large (about 1200), the elements of the vector $\mathbf{F}$ can be assumed to have an approximate joint normal distribution, as a consequence of the Central Limit Theory (Koch *et al.* 1977).

Weighted Least Squares Analysis

The ANOVA model for the transformed data $\mathbf{F} = \mathbf{Ap}$ obtainable from an experiment of a randomized complete block design can be written as:

$$\mathbf{F} = \mathbf{1}\mu + \mathbf{X}_B\beta + \mathbf{X}_I\tau + \mathbf{e}$$

where 1 is the $bv \times 1$ unit vector (of unit elements, where b and r are block and treatment number, repectively), $\mathbf{X}_B$ is the $bv \times b$ design matrix for blocks and $\mathbf{X}_I$ is the $bv \times v$ design matrix for treatments; where β is a $b \times 1$ vector of the block effects, τ is a $v \times 1$ vector of treatment effects and $\mathbf{e}$ is a $bv \times 1$ vector of random errors (environmental effects), with the expectation $E(\mathbf{e}) = \mathbf{0}$ and the variance-covariance matrix (dispersion matrix) $V(\mathbf{e})$, which will be denoted here by $\mathbf{V}_F$. More precisely, the elements of $\boldsymbol{\beta}$ and $\boldsymbol{\tau}$ represent the effects of blocks and of treatments, respectively, on the linear regression of the observed proportion (as dependent variable) on ear segment (as independent variable). Thus, $\boldsymbol{\tau}$ can be defined as the vector of the true linear effects of the interaction between lines and ear segments.

Taking into account the non-homogeneity of the variances, the linear model was fitted to the data obtained as $\mathbf{F} = \mathbf{Ap}$ by means of weighted least square analysis.

The appropriate matrix of weights for $\mathbf{F}$ is:

$$\mathbf{W}_F = [\mathbf{A}\, V(\mathbf{p})\, \mathbf{A}']^{-1} = \mathbf{V}_F^{-1},$$

with $V(\mathbf{p})$ as an $5bv \times 5bv$ block diagonal matrix having on the main diagonal the 5×5 matrices

$$V(\mathbf{p}_l) = n_l^{-1}\, (\mathbf{p}_l^\delta - \mathbf{p}_l \mathbf{p}_l'),$$

where $\mathbf{p}_l$ is a 5×1 vector of proportions observed at the 5 segments on the l-th experimental unit, $\mathbf{p}_l^\delta$ is a 5×5 diagonal matrix with the elements of $\mathbf{p}_l$ on the main diagonal, and $\mathbf{n}_l$ is the total number of observations on which $\mathbf{p}_l$ is based ($l = 1, 2, \ldots, bv$). It may be noticed that $\mathbf{V}_F$ is a $bv \times bv$ diagonal matrix, its diagonal elements being $\mathbf{a}'\, V(\mathbf{p}_l)\mathbf{a}$, where a is the 5×1 vector defined as

$$\mathbf{a} = [-2, -1, 0, 1, 2]'.$$

The working model is now

$$\tilde{\mathbf{F}} = \mathbf{W}_F^{1/2} F = \mathbf{W}_F^{1/2}\, \mathbf{1}\mu + \tilde{\mathbf{X}}_B \boldsymbol{\beta} + \tilde{\mathbf{X}}_T \boldsymbol{\tau} + \tilde{\mathbf{e}},$$

where

$$\tilde{\mathbf{X}}_B = \mathbf{W}_F^{1/2} \mathbf{X}_B,\ \tilde{\mathbf{X}}_T = \mathbf{W}_F^{1/2} \mathbf{X}_T \text{ and } \tilde{\mathbf{e}} = \mathbf{W}_F^{1/2} \mathbf{e},$$

with $E(\tilde{\mathbf{e}}) = \mathbf{0}$ and $V(\tilde{\mathbf{e}}) = \mathbf{I}$ (the $bv \times bv$ identity matrix).

The treatment sum of squares, adjusted for block effects, is (cf. Pearce 1983, section 3.3) of the form

$$\text{TSS} = \tilde{\mathbf{F}}'\, \boldsymbol{\Phi}\, \tilde{\mathbf{X}}_T (\tilde{\mathbf{X}}_T'\, \boldsymbol{\Phi}\, \tilde{\mathbf{X}}_T) - \tilde{\mathbf{X}}_T'\, \boldsymbol{\Phi}\, \tilde{\mathbf{F}} = \tilde{\mathbf{Q}}'\, \check{\mathbf{C}}^-\, \tilde{\mathbf{Q}},$$

where $\tilde{\mathbf{Q}} = \tilde{\mathbf{X}}_T'\, \tilde{\boldsymbol{\Phi}}\, \tilde{\mathbf{F}}$, while $\check{\mathbf{C}}^-$ is any generalized inverse of the coefficient matrix of the "reduced" normal equations for treatments, i. e. of $\check{\mathbf{C}} = \tilde{\mathbf{X}}_T' \tilde{\boldsymbol{\Phi}}\, \tilde{\mathbf{X}}_T$, with $\tilde{\boldsymbol{\Phi}} = \mathbf{I} - \tilde{\mathbf{X}}_B\, \tilde{\mathbf{k}}^{-\delta}\, \tilde{\mathbf{X}}_B'$ and $\tilde{\mathbf{k}}^{-\delta} = (\tilde{\mathbf{X}}_B' \tilde{\mathbf{X}}_B)^{-1}$, the latter being a diagonal matrix of the diagonal elements equal to the recipro-cals of $w_{.j} = \Sigma w_{ij}$, where w_{ij} denotes the weight corresponding to the i-th treatment in the j-th block.

The component of the treatment sum of squares (TSS) correspond-ing to the hypothesis $H_0: U'\tau = 0$ is of the form

$$TSS(U) = \tilde{F}'\tilde{\Phi}\,\tilde{X}_T\,\tilde{C}^-U(U'\tilde{C}^-U)^{-1}\,U\,\tilde{C}^-\,\tilde{X}'_T\,\tilde{\Phi}\,\tilde{F}$$

$$= \tilde{Q}'\tilde{C}^-U(U'\tilde{C}^-U)^{-1}\,U'\tilde{C}^-\tilde{Q},$$

which is applicable for any matrix U of v rows and of linearly inde-pendent columns, satisfying the estimability condition $U = \tilde{C}\,\tilde{C}^-U$.

Finally, the error sum of square is obtainable in the form

$$ESS = \tilde{F}'\tilde{\Phi}\,\tilde{F} - TSS$$

$$= \tilde{F}'\{\tilde{\Phi} - \tilde{\Phi}\,\tilde{X}_T\tilde{C}^-\tilde{X}'_T\tilde{\Phi}\}\,F = \tilde{F}'\,\tilde{\psi}\,\tilde{F},$$

say (cf. Pearce 1983, p.63).

The χ^2 criterion can be used for both, to test the goodness of fit of the model and to test the hypothesis H_0, as the statistics ESS and TSS(U) have asymptotically (central) χ^2-distributions, with degrees of freedom (DF) equal to tr $\tilde{\psi} = (b - 1)(v - 1)$ and to rank U, respec-tively, the distribution of TSS(U) being central if H_0 is true (see Grizzle, Starmer and Koch 1969).

Mean square expectation and heritability

The expectations for the sum of squares are obtained by the general formula for the expectation of a quadratic form (see Rao 1973, sections 2b.3 and 4a.1). In particular, the (conditional) expectation of TSS(U) for τ fixed can be written as:

$$E[TSS(U)|\tau] = \text{tr}\,\tilde{C}^-U(U'\tilde{C}^-U)^{-1}\,U\tilde{C}^-\,V(\tilde{Q})$$

$$+ E(\tilde{Q}')\,\tilde{C}^-U(U'\tilde{C}^-U)^{-1}\,U\,\tilde{C}^-\,E(\tilde{Q}),$$

which, due to the properties of $\tilde{Q}$, $E(\tilde{Q}) = \tilde{C}\tau$ and $V(\tilde{Q}) = \tilde{C}$, and to the conditions satisfied by U, reduces to:

$$E[TSS(U)|\tau] = \text{tr}(U'\tilde{C}^-U)^{-1}\,U'\tilde{C}^-U + \tau'U(U'\tilde{C}^-U)^{-1}\,U'\tau$$

$$= \text{rank}\,U + \tau'U(U'\tilde{C}^-U)^-\,U'\tau.$$

This can also be written (for computational convenience) as:

$$E[TSS(U)|\tau] = \text{rank } U + \tau'U\,M'M\,U'\tau,$$

if M is the inverse of the lower triangular Cholesky decomposition matrix of $U'\check{C}^-U$, giving $(U'\check{C}^-U)^{-1} = M'M$ (see Goodnight and Speed 1978).

Now, if τ is considered as an actual realization of a random vector T with a variance-covariance matrix $V(T)$ and the expectation $E(T) = 0$, then the unconditional expectation of $TSS(U)$ gets the form

$$E[TSS(U)] = E\{E[TSS(U)|T]\}$$
$$= \text{rank } U + \text{tr } U(U'C^-U)^{-1}U'V(T)$$
$$= \text{rank } U + \text{tr } M U'V(T)U M'.$$

As to the error sum of squares (ESS), its expectation is

$$E(ESS) = \text{tr } \tilde{\psi} V(\tilde{e}) + E(\tilde{e}') \tilde{\psi} E(\tilde{e}) = \text{tr } \tilde{\psi} = (b-1)(v-1),$$

since $\tilde{\psi}\tilde{F} = \tilde{e}$ (cf. Pearce 1983, p.63), $E(\tilde{e}) = 0$, $V(\tilde{e}) = I$, $\text{tr } \tilde{\Phi} = bv - \text{tr } \tilde{X}_B k^{-\delta} \tilde{X}'_B = bv - b$, and $\text{tr } \tilde{\Phi}\tilde{X}_T \check{C}^- \tilde{X}'_T\tilde{\Phi} = \text{rank } \check{C} = \text{rank } C = v - 1$. Hence, the expectation of the error mean square, $EMS = [(b-1)(v-1)]^{-1} ESS$, is, under the assumed model, equal to 1, i.e. $E(EMS) = 1$.

In the present experimental context, there are $v = 2t = 60$ treatments: $t = 30$ lines (families) obtained at a low selection intensity and $t = 30$ lines (families) obtained at a high selection intensity. This implies testing of two hypotheses. One involves the $2t \times 1$ vector

$$u_0 = [-1, -1, \ldots, -1, 1, 1, \ldots, 1]' = ([-1, 1] \otimes 1'_t)'$$

defining the contrast between groups (of selection intensity), where 1_t is the $t \times 1$ unit vector. Another hypothesis involves the $2t \times 2(t-1)$ matrix U_1 determining contrasts between lines (families) within groups, independent of the contrast between the groups. It can be defined by the following conditions:

$$\text{rank } U_1 = 2(t-1), \qquad U'_1 \begin{bmatrix} 1_t & 0 \\ 0 & 1_t \end{bmatrix} = 0.$$

Hence, the relevant components of the treatment sum of squares are

$$TSS(u_0) = \tilde{Q}'\check{C}^-u_0(u_0'\check{C}^-u_0)^{-1} u_0'\check{C}^-\tilde{Q} = (u_0'\check{C}^-\tilde{Q})^2/u_0'\check{C}^-u_0$$

and

$$\text{TSS}(\mathbf{U}_1) = \tilde{\mathbf{Q}}' \tilde{\mathbf{C}}^- \mathbf{U}_1 (\mathbf{U}_1' \tilde{\mathbf{C}}^- \mathbf{U}_1)^{-1} \mathbf{U}_1' \tilde{\mathbf{C}}^- \tilde{\mathbf{Q}},$$

invariantly on the choice of C^- and also on the choice of the matrix U_1, satisfying the above conditions.

To obtain the expectations of these sums of squares, the matrix $V(T)$ is to be specified. A reasonable (the usual) assumption is that:

$$V(\mathbf{T}) = \sigma^2_{F(G)}(\mathbf{I}_2 \otimes \mathbf{I}_t) + \sigma^2_G(\mathbf{I}_2 \otimes \mathbf{1}_t \mathbf{1}_t'),$$

where $\sigma^2_{F(G)}$ is the variance component for families (lines) nested within the groups, and σ^2_G is the variance component for the groups. Under this assumption,

$$
\begin{aligned}
E[\text{TSS}(\mathbf{u}_0)] &= \text{rank } \mathbf{u}_0 + \text{tr } \mathbf{m}_0 \mathbf{u}_0' V(\mathbf{T}) \mathbf{u}_0 \mathbf{m}_0' \\
&= 1 + 2\mathbf{m}_0' \mathbf{m}_0 (t\sigma^2_{F(G)} + t^2 \sigma^2_G) \\
&= 1 + 2t(\mathbf{u}_0' \tilde{\mathbf{C}}^- \mathbf{u}_0)^{-1} (\sigma^2_{F(G)} + t\sigma^2_G)
\end{aligned}
$$

and

$$
\begin{aligned}
E[\text{TSS}(\mathbf{U}_1)] &= \text{rank } \mathbf{U}_1 + \text{tr } \mathbf{M}_1 \mathbf{U}_1' V(\mathbf{T}) \mathbf{U}_1 \mathbf{M}_1' \\
&= 2(t-1) + \text{tr } \mathbf{M}_1 \mathbf{U}_1' \mathbf{U}_1 \mathbf{M}_1' \sigma^2_{F(G)} \\
&= 2(t-1) + \text{tr } (\mathbf{U}_1' \tilde{\mathbf{C}}^- \mathbf{U}_1)^{-1} \mathbf{U}_1' \mathbf{U}_1 \sigma^2_{F(G)} \\
&= 2(t-1) + \text{tr } (\mathbf{U}_1' \tilde{\mathbf{C}}^- \mathbf{U}_1)^{-1} \sigma^2_{F(G)},
\end{aligned}
$$

the last form in the case where $\mathbf{U}_1$ is chosen so as to have orthonormal columns.

The resulting expectations can, therefore, be written as in Table 2.

The computing procedure for GMS and F(G)MS expectations corresponds to that adopted in the SAS System program (GLM

TABLE 2

Source	MS	Expected mean squares
Groups	GMS	$1 + 2t(\mathbf{u}_0' \tilde{\mathbf{C}}^- \mathbf{u}_0)^{-1} \sigma^2_{F(G)} + 2t^2(\mathbf{u}_0' \tilde{\mathbf{C}}^- \mathbf{u}_0)^{-1} \sigma^2_G$
Families (within groups)	F(G)MS	$1 + [2(t-1)]^{-1} \text{tr} (\mathbf{U}_1' \tilde{\mathbf{C}}^{-1} \mathbf{U}_1)^{-1} \mathbf{U}_1' \mathbf{U}_1 \sigma^2_{F(G)}$
Error	EMS	1

Procedure, 'Expected Mean Square for Random Effects'). Therefore this program was used for MS and variance coefficient evaluation.

The proportion of genetic variability between families (h_F^2) can be obtained as:

$$h_F^2 = \frac{\sigma_G^2 + \sigma_{F(G)}^2}{\sigma_G^2 + \sigma_{F(G)}^2 + \sigma_e^2}$$

and its component solely due to gametophytic control as

$$h_G^2 = \frac{\sigma_G^2}{\sigma_G^2 + \sigma_{F(G)}^2 + \sigma_e^2},$$

where $\sigma_e^2 = 1$

Standard Errors of variance components were obtained according to the procedure given by Searle (1971) for balanced data. Therefore, these estimates are to be considered approximate values. The expected mean squares, the proportion of genetic variability between families and the standard errors for sporophytic traits were computed by standard methods (Becker 1975).

RESULTS AND CONCLUSIONS

The numerical example is based on the data produced by Ottaviano, Sari Gorla and Villa (1988). The values of the coefficients in the three expectations and the estimated values of the variance components are reported in Table 3, together with their standard errors (procedure given by Searle 1971).

TABLE 3
Anova of pollen trait: tube growth rate

Source	SS	DF	χ^2	Expected mean squares	$\hat{\sigma}^2$	SE
G	103.41	1	**	$1 + 1.82\,\sigma_{F(G)}^2 + 54.58\,\sigma_G^2$	1.68	1.55
$F(G)$	812.98	58	**	$1 + 2.20\,\sigma_{F(G)}^2$	5.92	1.16
Error	1.82	59	ns	1	1	0.01

ns: not significant.
**: $p < 0.01$.

For the same families sporophytic traits were also studied (Table 4). This is a straightforward ANOVA for quantitative traits, for which normality of distribution and homogeneity of variances can be assumed. The estimates of variance components and of h^2 have been obtained as in the previous section.

The estimated proportions of the population variance describing the phenomenon are reported in Table 5. The variance released by MGS is the proportion of population variance due to genes expressed in the gametophytic phase. It is a direct proof in favour of the first biological assumption: pollen population shows genetical variability in fitness. Variance of sporophytic traits released by gametophytic selection proves the second assumption: MGS affects also the sporophytic (plant) population.

TABLE 4
Anova of plant trait: 50 kernel weight

Source	DF	MS	EMS	$\hat{\sigma}^2$	SE
Blocks	2	8.12			
Groups	1	157.14**	$\sigma_e^2 + 3\sigma_{F(G)}^2 + 240\sigma_G^2$	0.62	0.53
Full-sib families (within groups)	158	7.48**	$\sigma_e^2 + 3\sigma_{F(G)}^2$	1.95	0.28
Error	318	1.647	σ_e^2	1.65	0.13

TABLE 5
Proportions of phenotypic variability

Pollen tube growth rate	Environment (0.12)	
	Genetics (0.88)	Released by MGS (0.19)
		Residual (0.69)
Kernel weight	Environment (0.39)	
	Genetics (0.61)	Released by MGS (0.15)
		Residual (0.46)

It is worth while to note that the proportion of pollen fitness variability controlled by the haploid genome is quite high: the variance released by MGS is probably underestimated, since only a few generations of selection should not exhaust the existing potential variability. The response to selection obtained in the sporophyte generation indicates that genes controlling important physiological processes and conferring adaptive value, are common to both phases of the life cycle (are expressed both in pollen and plant).

The results of this study strongly support the significance of MGS in the evolution of higher plants: it could represent a basic mechanism regulating polymorphism and genetic load and could account for a high evolution rate.

With regard to the mechanism regulating the genetic load, we mainly refer to quantitative traits, which are generally determined by complex gene combinations. For these characters a large amount of load is expected to be produced in each generation by genetic recombination. Because of the large population size, selection acting on the gametophytic generation can remove a large portion of the recombination products, at a cost which is compatible with the biological feature of the species. As mentioned above, the same argument can justify a high evolution rate. Indeed, at a high intensity of selection and for deleterious recessive genes, the time required for a significant reduction of gene frequencies is much greater under diploid selection then under haploid selection (Ottaviano and Sari Gorla 1979).

As far as the statistical procedure is concerned, it is important to note that it can be a very efficient tool for several experimental situations, where the evaluation of genetic variability and response to selection is concerned. In fact, several important traits related to fitness and of economic importance, such as disease resistance, show binomial distribution and vary according to the developmental stage; in these cases most of the statistical approaches generally used offer only approximate solutions, since they are based on assumptions that are not fully satisfied by the experimental data.

° *Academy of Agriculture in Poznań, Poland*
* *Università di Milano, Italy*

BIBLIOGRAPHY

Becker, W. A. 1975. *Manual of quantitative genetics*. Washington State University Press, Pullman: Washington State University Press.

Fisher, R. A. 1930. *The genetical theory of natural selection*. Oxford: Clavendon Press.

Goodnight, J. H. and F. M. Speed. 1978. *Computing expected mean squares*, Cary, N.C: SAS Institute Inc.

Grizzle, J. E., Starmer, C. F. and G. G. Koch. 1969. Analysis of categorical data by linear models, *Biometrics* **25** 489—504.

Koch, G. G., Landis, J. R., Freeman, J. L. and D. H. Freeman. 1977. A general methodology for the analysis of experiments with repeated measurement of categorical data. *Biometrics* **33** 133—158.

Ottaviano, E. and M. Sari Gorla. 1979. Genetic variability of male gametophyte in maize. Pollen genotype and pollen-style interaction. Israeli-Italian joint meeting on Genetics and Breeding of Crop Plants. *Monogr Genet Agr*: 89—106.

Ottaviano, E., Sari Gorla, M. and M. Villa. 1988. Components of the male gametophytic fitness in maize. Genetic variability and correlation with sporophytic traits. *Theor Appl Genet* (to appear).

Pearce, S. C. (1983). *The agricultural field experiment*, Chichester and New York: John Wiley & Sons.

Rao, C. R. 1973. *Linear statistical inference and its applications*. New York: John Wiley & Sons.

Searle, S. R. 1971. *Linear models*. New York: John Wiley & Sons.

ALBERTO PIAZZA

STATISTICS IN GENETICS: HUMAN MIGRATIONS DETECTED BY MULTIVARIATE TECHNIQUES

INTRODUCTION

The role of migration as a mechanism influencing the genetic structure of human populations has long been recognized. Human migrations, however, can occur in several different ways, each with its characteristic pattern in time and space. An exhaustive analysis of this evolutionary pressure, how it interacts with biological factors as natural selection and random genetic drift, and with no biological constraints like geographic distances and barriers cannot be condensed in few pages. We rather focus on empirical studies and specifically on how human migration can be detected by the multivariate technique of principal component analysis.

Empirical studies mostly refer to *gene frequency* data. All blood groups with the important addition of proteins and enzymes whose variations can be detected by electrophoretic techniques are genetically controlled markers. Gene frequencies usually differ in different populations.

The main problem to discuss is whether and in what proportion migration can explain the gene frequency geographic variation we observe in human populations.

It is convenient to distinguish three relevant aspects of the matter:

a) how migration compares with other mechanisms of evolution in causing genetic differences;
b) how to estimate these genetic differences; and
c) how these differences allow to discriminate migration from other evolutionary factors.

It may be useful to enumerate at least three different kinds of migration in the human context. Depending on the ratio: number of people who move to a certain geographical area over the number of people already settled in that area, we refer to processes of *individual* or *massive* migrations. They are associated to the micro- or macro-geographic structures of our species. Furthermore in the past history of

R. Cooke and D. Costantini (eds), Statistics in Science. The Foundations of Statistical Methods in Biology, Physics and Economics, 103–118.
© 1990 Kluwer Academic Publishers.

Man massive migrations have often been accompanied by population expansions giving rise to continuous waves of advance of people and genes, and this cumulative effect of migration with population growth has been called 'demic expansion' (Wijsman and Cavalli-Sforza, 1984).

MIGRATION AND OTHER MECHANISMS OF EVOLUTION

In addition to migration, mutation, natural selection and random genetic drift are the major forces of evolution.

All existing alleles originate through mutations, but the frequency with which they occur seems to be rarely if ever above 1 per 10 000 gametes per generation per locus. In the genetic differentiation of our species, *Homo sapiens sapiens*, which occupies at most 50 000— 100 000 years, mutation rates alone are likely to play a secondary role, if any. However when multiple mutations at the same locus are associated to massive movements of people, their geographical distribution may suggest the migration site of origin and possibly its direction.

An interesting example is the observation that the frequency distribution of three variants of human haemoglobin gene (Pagnier *et al.*, 1984) is compatible with the Bantu expansion having originated about the time of the Roman Empire collapse in an area close to the present day eastern Nigeria.

Natural selection has the distinction of being not only the unique adaptive mechanism of evolution, but also the fastest. Assuming a simple model of selection where the 'favoured' genetic type has a fitness 10% greater than the 'normal' one, the time necessary for the gene frequency of the favoured type to increase say from 10 to 90% is about 30 generations, thus well in the span of life of our species and perfectly compatible with a process of migration giving the same amount of differentiation in the same time. However natural selection is going to affect every gene in a different way and differently in different environments, while migration affects all genes in a similar way.

This last consideration would lead to statistical methodologies allowing to distinguish genetic differentiation by migration from genetic differentiation by natural selection, if random genetic drift, the stochastic component of genetic transmission, does not play any substantial role in human evolution. However — specially in certain periods of our history and in special geographic areas — populations are structured in samples of small sizes and random fluctuations of their gene frequencies taking place at every generation — the evolutionary result produced by *genetic drift* — are indeed the rule rather than the exception. Also random genetic drift affects all genes equally, being entirely dependent on the sample sizes of the population and not on the single gene. An empirical way to discriminate genetic drift from migration is by observing that in a population the average variability in gene frequencies usually increases with migration from outside and decreases with genetic drift. In fact a sort of equilibrium between the two pressures may occur that could explain why a genetic differentiation can be kept stable in time.

ON DETECTING MASS MIGRATIONS AND THEIR SITE OF ORIGIN

Remote and recent history has seen many *mass migrations*, usually from areas of high population growth to areas considered desirable for some reasons (e.g. for being particularly wealthy, economically attractive, etc.).

Suppose that a population migrates from one place to another already inhabited by people who are distinct from the immigrants in the frequencies of several genes. The resulting admixture may give rise to a mixed population whose gene frequencies will be the weighted averages of the gene frequencies of the parental populations. The weights are given by the proportion of the parental populations in the admixture, and will, therefore, be equal for all genes.

Mass migrations thus may determine linear transformations of gene frequencies. It seems reasonable that the linear procedures of multivariate analysis can be especially useful in the examination of the gene frequencies which have been altered by these linear processes. We have employed *principal components*, which aptly summarize all the information shared by many genes in the area being studied. The hypothesis is that each principal component, because of reciprocal orthogonality, can possibly detect independent mass migrations that occurred at different times with different histories.

In the more recent prehistory of Man there have been several population expansions over areas previously unoccupied (such as New Guinea, Australia, the Pacific Ocean islands and many others) or inhabited by other populations with which admixture may or may not have occurred, as for instance the Bantu spread to Central and South Africa.

We have been especially interested in the possibility of explaining the occupation of Europe by neolithic farmers originating in the Middle East and mixing with preexisting mesolithic, i.e. latest European paleolithic hunter-gatherers. To this aim archeological data bearing on the advance of agriculture in Europe have been analysed by Ammerman and Cavalli-Sforza (1971).

Some time later genetic data on modern European populations have been examined using contour maps of the principal components of gene frequencies (Menozzi et al., 1978). They showed excellent agreement with the archaeological observations in supporting the hypothesis of a migration from the Middle East, and in addition indicated the possible existence of more than one migration.

THE POWER OF PRINCIPAL COMPONENT IN SEPARATING DIFFERENT WAVES OF EXPANSION: A SIMULATION

To test the possibility of separating independent migrations, the spread over Europe of populations coming from different places at different times has been simulated (Rendine et al., 1986). Europe and Middle East were represented in the simulation as a number of squares, each of the average size expected for a 'tribe' of hunter-gatherers: such elementary areas were of (156×156) km^2 each, and the whole territory of interest corresponded to a total of 24×35 elementary areas, including seas and mountainous areas which were left unoccupied.

The region of initial neolithic development was placed in the traditional area of origin of the Middle East, by connecting 6 adjacent tribes in the region corresponding approximately to present Israel and Lebanon and giving them higher values for the maximum population density and for migration rates. This gave rise to the 'farmers' community originated in the Middle East, which from the onset of agriculture was kept distinct from the original hunter-gatherers and allowed to migrate also outside their original area.

Genes from 20 diallelic loci were started at a frequency of 50%. Migration, genetic drift and different population growth for farmers and hunter-gatherers have been allowed to evolve for about 10 000 years in order to simulate the distribution of the present-day European gene frequencies.

As we were interested in testing the power of principal components in separating different waves of expansion, new expansions have been introduced after the first due to agriculture starting from the Middle East. We took in fact as origins of new expansions, areas which had been recognized as maxima (or minima) in the surface of the second and third principal components of European gene frequencies (Menozzi *et al.*, 1978).

One such area in the plot of the second principal component is located at the North Eastern boundary of Europe, and was interpreted as due to the migrations of several peoples from North-West Asia to Europe which were known to have taken place in the last 3000 years. Another area was centred around the Black Sea and corresponded to the third principal component: it might be interpreted either as the Indo-European expansion, which is assumed by some linguists and anthropologists to have started from an area approximately corresponding to this one more than 5000 years ago.

At the end of the process (10 000 years), principal components of the 20 gene frequencies were calculated for each elementary area, and those corresponding to the highest 4 eigenvalues were plotted as contour maps (Rendine *et al.*, 1986). The first principal component showed the expected pattern, similar to the observed first principal component of real data and corresponding to the geographic pattern of advance of farming itself from the Middle East, both simulated and real.

The second principal component had a centre in North Africa perhaps in response to the Mediterranean barrier.

The third principal component produced a pattern in which there is an apparent centre of expansion from the North East located near the Black Sea which corresponds to the centre of origin of the second wave of expansion.

The fourth principal component showed a possible centre of origin in the area of the third wave of expansion (N.E. Asia). Clearly the second and the third waves of expansion were too close to be distinguished as neatly as the first one.

In spite of the need to have used simulations of enormous simplicitly compared with actual situations, our work suggests that:

a) a pace of migration and growth similar to that found in appropriate ethnographic situations is well compatible with the rate of demic diffusion of neolithic farmers;
b) the rate of transformation and the increase of individuals allowed by

new cultural innovations followed by expansion are critical parameters in determining patterns of geographic distribution of genes, their clines, and the separability by principal component analysis of independent diffusions;

c) when the diffusion of people has reached its limits and the carrying capacity of the involved area is saturated, the genetic gradients thus reached change very slowly thereafter.

The last point is somewhat surprising, since one would expect that migration should level off any gradient that is not maintained by selection, but at least in the one-dimensional case it can be proved by simple algebra.

The qualitative finding that the process of equalization of genetic gradients can be extremely slow, supports the idea of a very long memory of the observed genetic gradients and it has the practical implication of giving a methodological motivation to our efforts of recognizing migration processes by the principal component analysis of gene frequencies.

SHORT-RANGE MIGRATIONS IN ITALY

Over the last century, with the industrial revolution and the corresponding concentration of people in very large urban settlements, there were massive migrations and great changes in the population structure of most European countries. Italy, in particular, has changed entirely over the last thirty years, as the concentration of industries in the north induced an internal south-north migration of unexpected proportions (Golini, 1974).

The slow pace of genetic changes and the relatively gross measurements of the traditional blood groups cannot resolve this process. Depth in time is too small for a genetic difference to be singled out from sampling noise when using a gene such as RH or ABO or any of the others made available to the geneticist. The power of these genes is limited by the number of 'variant' types: the higher the number of variants (or *alleles*) of a gene, the higher the potential number of genetically different individuals and the more powerful this gene can be for resolving small differences in the geographical distribution of its frequencies. A genetic system that is inexpensive, simple to test and sensitive enough to detect short-range migrations, is that of surnames. Surnames, considered as alleles of a gene transmitted only by the male line, can be assumed to be genetic markers unaffected by natural selection (Zei et al., 1983) and therefore satisfy the expectations of a theory of evolution based only on random genetic drift, mutation and migration (Karlin and McGregor, 1967; Ewens, 1972). By making use of this theory we compared the estimates of migration rates in Italy, as inferred by the surname distribution found in the telephone directories, with the corresponding estimates from official demographic sources (Piazza et al., 1987). It was a nice surprise to find that at least with this source of data, the ratio of surnames to individuals makes it possible to obtain reliable estimates of migration rates.

If N is the total number of individuals who form a sample and S is the number of surnames found in that sample, the following system of two equations:

(1a) $\quad N = \alpha(1 - v)/v$

(1b) $\quad S = -\alpha \log v$

allows v, the immigration rate, and the parameter α to be obtained given N and S (Fisher, 1943; Zei et al., 1983).

Surname data were collected from the telephone directories of 91 Italian provinces after elimination of commercial subscribers. Our sample has 10 473 727 registered telephone users with 59 961 000 inhabitants, i.e. 1 telephone every 5.44 residents: therefore most families are represented. Immigration was estimated from (1a), rewritten as:

(2) $\quad v = \alpha/(N_m + \alpha)$.

The parameter α was taken from the solution of system (1) with S and N being the number of surnames and the number of telephone users provided for each province of Italy by the telephone directories, N_m is the *male* census size from the same provinces. The immigration rates v, as estimated by (2), have been compared with actual immigration rates, calculated by averaging the yearly ratios 1967—1980 of newly registered individuals to total residents in the 91 Italian provinces, as published by the Italian registration offices (Istat, 1967—1980).

The correlation between 'observed' and 'estimated' migration rates is 0.596 ± 0.083, significantly different from zero. Special statistical techniques have been used to detect possible outliers. Of these putative outliers, the most extreme (Trieste) was only recently annexed to Italy; history and tourism may be influential for Venezia and Genova which deviate in the same direction as Trieste and may be responsible for an anomalous pattern of migration. The interpretation of the other outliers corresponding to the towns of Nuoro, Caserta, Avellino is still to be investigated: all three belong to regions (Sardinia and Campania) where emigration was remarkably higher than immigration in the considered period of time.

In conclusion the surnames, considered as alleles of an extremely polymorphic genetic marker behaving as linked to the Y chromosome, seem to provide a very useful tool for studying short-range migrations when other demographic sources are not available.

The unpleasant side of the story, however, is that surnames describe a sexist evolution, the migration of the males being the only one considered. Furthermore their introduction is too recent for tracing old, long-range movements and settlements of populations: more appropriate albeit less direct approaches must be developed.

LARGE-SCALE MIGRATIONS IN ITALY

An analysis we made for identifying putative migrations into Italy possibly older than those detected by the evolution of surnames has already been used to interpret the geographical distribution of the gene frequencies in Europe (Menozzi et al., 1978) and in the world (Piazza et al., 1981). The multivariate technique of principal components has been applied. As explained above, the synthetic displays provided by plotting the principal component scores are particularly useful in detecting gradients (clines) of genetic differentiation associated with movements of populations like those accompanying the Neolithic expansion of farmers from the Near East (Rendine et al., 1986) or, in more recent times, the putative diffusion of Indo-European speaking populations (Ammerman and Cavalli-Sforza, 1984; Renfrew, 1987).

We considered blood group gene frequencies quoted in more than 500 references for a total of 34 alleles mostly homogeneously distributed all over Italy; the results of the analysis can be summarized in synthetic images (Piazza et al., 1988) showing clines of genetic variation which we would like to discuss in some detail.

One of these images is represented in Figure 1. It is the plot of the

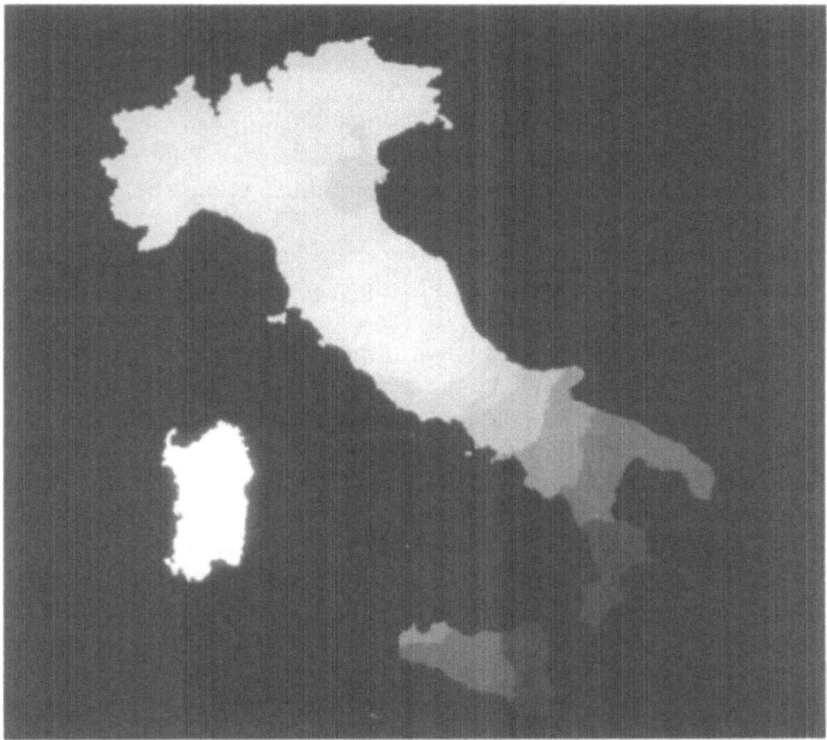

Fig. 1. Contour plot of the first principal component coordinates of genes frequencies in Italy from 34 independent alleles at the human loci: ABO, Rhesus, MNS, KELL, Haptoglobin, HLA-A, HLA-B. Shades indicate different values of the principal component scores. (from Piazza *et al.* 1988).

first principal component of all genetic data and it conveys 35% of the total variation. The island of Sardinia is not included in the map because its great genetic distance from the peninsula would flatten the genetic differentiation within Italy to a scale too small to be resolved by the human eye. The genetic isolation of Sardinia, is supported by many other pieces of evidence (Menozzi *et al.*, 1978; Olivetti *et al.*, 1986; Piazza *et al.*, 1985).

The Italic world, with its local cultures, seems to be more or less clearly defined at the beginning of the Iron age, i.e. since the IX—VIII century B.C., when an approximate correspondence between archaeologically defined cultures and linguistic areas has been established through epigraphic records and literary sources (Pallottino, 1984). Figure 2 shows the geographical distribution of languages in Italy as documented in the 5th century B.C. (Pulgram, 1958). Interestingly enough, such a pattern of language differentiation can still be recognized in the 'regions' and dialects of contemporary Italy and can be used as a reference to describe the various ethnic groups which formed Italy at the beginning of its history.

Literary tradition and archaeological evidence report a sharp increase in population in the eastern Mediterranean in the late part of the second millennium B.C. (McEvedy and Jones, 1978): a logical consequence of this was the foundation of new cities overseas ('colonies'). During the eighth century B.C. the western Mediterranean was the site of a complex network of colonies founded mainly by Phoenicians and Greeks (Sherrat, 1980). Whereas the Phoenicians directed their main colonizing efforts towards the coasts of North Africa, Spain, Malta, Sardinia and the western triangle of Sicily, the Greeks settled mainly along the southern and western shores of the mainland and also along the fertile coastal belt of Sicily (excluding the Phoenician western triangle). More than forty towns are known whose size and architectual magnificence underline the importance of what came to be known as *Magna Graecia*, Greater Greece: its geographical area of influence is shown in Figure 2.

An interesting feature of our genetic map (Figure 1) displaying the first principal component of 34 genes, is a clear North-South gradient showing a remarkable differentiation of *Magna Graecia* from the rest of the mainland. The genetic similarity of the Italian southern regions with modern Greece as shown by their similar gene frequencies, the genetic difference between the two parts of Sicily which appears in this map (the Phoenician colony of Motya in the western area of the island is very well defined: a matter of chance?) and finally the definition of a geographical area around the delta of the Po river which was described as the possible Adriatic site (Spina, Adria) of later (IV century B.C.) Greek settlements (Bérard, 1957) all suggest a Greek gene flow as the most relevant factor which could explain the genetic differentiation shown in this map of Italy. A non trivial question to be raised in order to make this interpretation more plausible is whether the Greek colonies were of such size to justify a diffusion of their genes. Because of the geographical position of Greece as the door from the Near East to the Mediterranean, by the end of the Bronze age (1000 B.C.) the

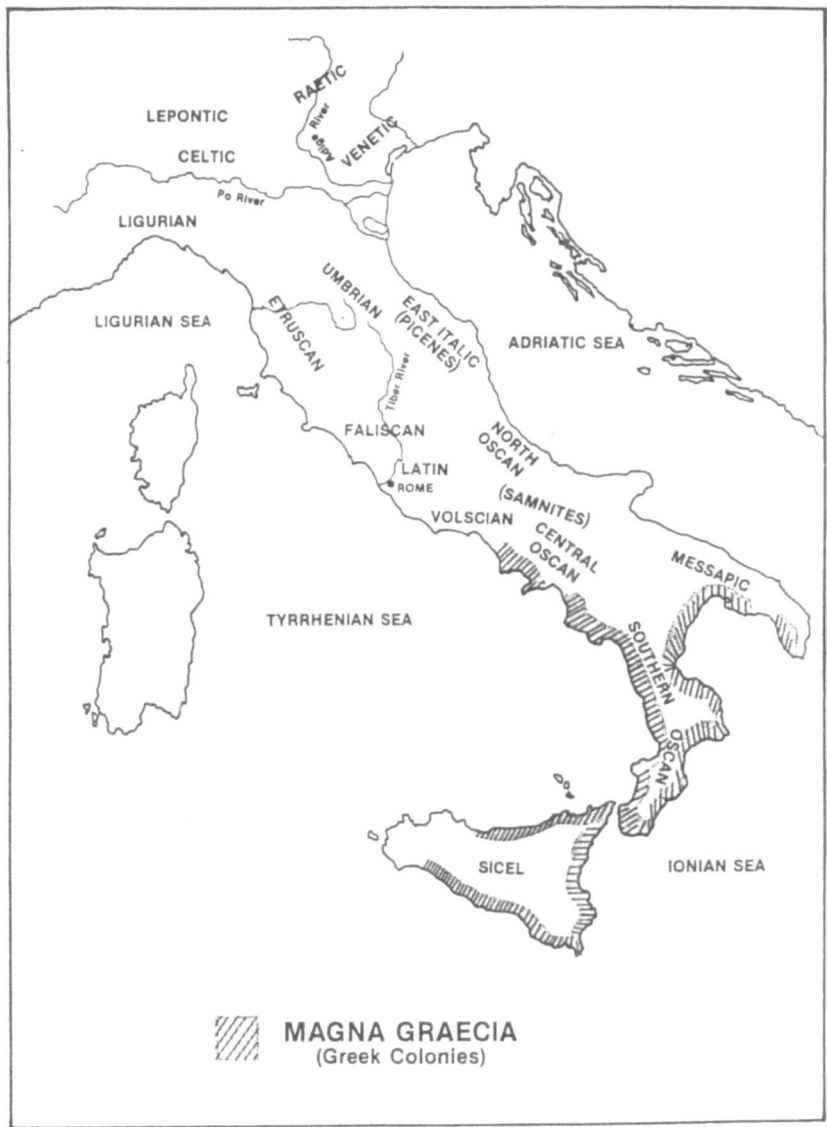

Fig. 2. The languages of ancient Italy at the end of the sixth century B.C. (adapted from Pulgram, 1958). The areas of the Greek colonization ('Magna Graecia', Greater Greece) are also shown.

average density of the population was higher in Greece than in Europe (3.7 inhabitants km^{-2}) by a factor of 3 (McEvedy and Jones, 1978). Between 1000 B.C. and 400 B.C. the population doubled in Europe, increasing from 10 to 20 million: in the same period the population trebled in Greece, reaching a total of 3 million (McEvedy and Jones, 1978). By 400 B.C. Italy, the second most densely populated country in Europe after Greece, had about 4 million people (McEvedy and Jones, 1978; Beloch, 1886). The Greek colonies of Sicily alone accounted for 1.5 million people, of which more than 10% (about 200 000) were of Greek origin (Beloch, 1886). To these Greek inhabitants of Sicily may be added at least another 100 000 Greek colonizers in the Italian peninsula, so that before the Roman period, one out of every 10—13 inhabitants in Italy was Greek (Beloch, 1886).

Even if these numbers must be taken with caution, their order of magnitude does not contradict the idea of a possible introduction in Sicily and in the South of Italy of Greek genes whose dilution in the Italian genetic pool is still apparent as a gradual genetic change from south to north.

The maps of the other principal components are discussed in the original paper (Piazza *et al.*, 1988): they also can be associated to putative settlements of pre-roman times, even if the sizes and the movements of the people involved remain much more obscure.

The lack of relevant archaeological records makes dating prehistoric movements of peoples and ascertaining their extent in Italy difficult. Linguistic records are sometimes the only source of useful information for identifying the different ethnic groups of prehistoric times. A main finding of our study is that languages and genes, at least as far as Italy is concerned, have a similar geographical distribution and therefore a probable common history. Comparison of linguistic records and archaeological cultures (mostly burial patterns) makes it possible to date the presence of the ancestors of the peoples mentioned in Figure 2 after the diffusion of the Indo-European languages and before or at the Early Iron Age (Peroni, 1979).

It seems plausible that the Indo-European languages were not brought to Italy through massive migrations, but rather by a moderate number of carriers originating roughly in the Danubian area, who reached Italy with metallurgical skills by a transalpine and/or an Adriatic route (Pulgram, 1958; Devoto, 1962).

The use of a language by a population is a cultural trait which can be imposed by a few politically powerful people without any substantial effect on the genetic structure of the population itself. However when it is associated to a specific genetic identity, then language and population may share a common origin. In the case of Italy, we suggest the preroman times as those of the main settlements and movements of peoples which determined the present pattern of genetic differentiation. Such early times are not as surprising as it must appear. In fact it has been shown that the genetic structure of the present European populations is likely to be the result of the expansion of Neolithic farmers from the Near East which occurred 10 000 years ago (Menozzi *et al.*, 1978; Ammerman and Cavalli Sforza, 1984). Moreover the computer simulation of the same process Rendine described above suggested that the clines of gene frequencies generated by this expansion are not so easily dissolved by successive migrations between contiguous populations once the expansion is over.

The romanization of Italy and Europe, which was obviously of the greatest importance for many other aspects of our history, is not likely to have changed the genetic individuality of the conquered populations in a substantial way. Colonization produced changes in the political, administrative, urban, commercial systems rather than a massive substitution of people such as occurred in the European colonization of the Americas or in the genocides of our times. A unique language, Latin, was also imposed by the Roman conquerors, but its adoption was not complete since even today each region in Italy speaks a different dialect revealing possible traces of the ancient language spoken by the ancestors of its inhabitants. A parallel analysis of the geographical distribution of gene frequencies in France (Piazza, 1986) seems to lead to similar conclusions as if the genetic roots of France were to be found before the settling of the Franks.

Our analysis provides information on the evolutionary history of the population — or rather the populations — of Italy. It is clear, however, that genetic similarities among populations living in different geographical areas might be due to environmental similarity or common origin, the corresponding differences being the outcome of natural selection or genetic drift balanced by migration. The study of the principal component gene frequency maps does not by itself allow a clear-cut choice between these possibilities, but natural selection as the main cause of the genetic gradients shown above could reasonably be excluded because of the following reasons:

a) genes clearly correlated with environmental or pathological factors, such as thalassemia, G-6-PD, etc., were excluded from the analysis;
b) the small geographical area of Italy and its temperate climate make significant changes in the physical environment implausible;
c) finally and most importantly, the representation of the principal components is by itself a technique which cannot easily accommodate selective agents:

in fact it displays the genetic differences shared by *all* genes, and it is very difficult to imagine a selective factor affecting a large fraction of genes at the same time and in the same proportion. On the contrary this holds in the case of genetic drift and migration: they depend on the population structure, i.e. on demographic parameters like size, rate of migration, growth, etc., but they do not discriminate among genes.

On the basis of these negative arguments, we suggest that the genetic history of Italy is not a classification of different kinds of environment, but more likely the result of the ancient history of peoples with their settlements and their movements. Genetic drift (i.e. stochastic effects due to the sample size) was the probable cause of differentiation among the many different ethnic groups who settled in that country. The migrational network established among them over the centuries contributed to originate the genetic gradients still discernable in our multivariate maps of modern Italy.

The extent to which these findings may provide statisticians with new insights about their science is probably not remarkable, but I am confident this application of standard multivariate techniques shows how fertile the statistical approach can be in other sciences: human genetics is an example.

Università di Torino, 10126 Torino, Italy

REFERENCES

Ammerman, A. J. and Cavalli-Sforza, L. L. 1971. 'Measuring the Rate of Spread of Early Farming in Europe'. *Man* **6** 674—688.
Ammerman, A. J. and Cavalli-Sforza, L. L. 1984. *The Neolithic Transition and the Genetics of Populations in Europe*. Princeton: Princeton University Press.
Beloch, J. 1886. *Die Bevölkerung der grieschisch-römischen Welt*. Leipzig.

Berard, J. 1957. *La colonisation grecque de l'Italie méridionale et de la Sicile dans l'antiquité*. Paris: Presses Universitaires de France.

Devoto, G. 1962. *Origini indoeuropee*. Firenze: Sansoni.

Ewens, W. J. 1972. 'The Sampling Theory of Selectively Neutral Alleles'. *Theor. Pop. Biol.* **3** 87—112.

Fisher, R. A. 1943. 'The Relation Between the Number of Species and the Number of Individuals in a Random Sample of an Animal Population'. *J. Anim. Ecol.* **12**, 42—58.

Golini, A. 1974. *Distribuzione della popolazione, migrazioni interne ed urbanizzazione in Italia*. Roma: Istituto Demografia Università.

Karlin, S. and McGregor, J. 1967. 'The Number of Mutant Forms Maintained in a Population'. *Proc. Fifth Berkeley Symp. Math. Sta. Prob.* **4** 415—438.

McEvedy, C. and Jones, R. 1978. *Atlas of World Population History*. Harmondsworth: Penguin.

Menozzi P., Piazza, A., and Cavalli-Sforza, L. L. 1978. 'Synthetic Maps of Human Gene Frequencies in Europeans'. *Science* **201** 786—792.

Olivetti, E., Rendine, S., Cappello, N., Curtoni, E. S., and Piazza, A. 1986. 'The HLA System in Italy'. *Hum. Hered.* **36** 357—372.

Pagnier, J., Mears, J. G., Dunda-Belkhodja, O., Schaefer-Rego, K. E., Beldjord, C., Nagel, R. L., and Labie, D. 1984. 'Evidence for the Multicentric Origin of the Sickle Cell Haemoglobin Gene in Africa'. *Proc. Natl. Acad. Sci. U.S.A.* **81** 1771—1773.

Pallottino, M. 1984. *Storia della prima Italia*. Milano: Rusconi.

Peroni, R. 1979. 'The Iron Age, Orientalizing and Etruscan Periods'. In *Italy before the Romans*, ed. by Ridgway, D. and Ridgway, F. R., pp. 7—30. London: Academic Press.

Piazza, A. 1986. 'The Genetic Data from the French Provinces: A Tentative Summary'. In *Human Population Genetics*, ed. by Ohayon, E. and Cambon-Thomsen, A., pp. 345—352. Paris: Inserm.

Piazza, A., Cappello, N., Olivetti, E., and Rendine, S. 1988. 'A Genetic History of Italy'. *Ann. Hum. Genet.* **52** 203—213.

Piazza, A., Mayr, W. R., Contu, L., Amoroso, A., Borelli, I., Curtoni, E. S., Marcello, C., Moroni, A., Olivetti, E., Richiardi, P., and Ceppellini, R. 1985. 'Genetic and Population Structure of Four Sardinian Villages'. *Ann. Hum. Genet.* **49**, 47—63.

Piazza, A., Menozzi, P., and Cavalli-Sforza, L. L. 1981. 'Synthetic Gene Frequency Maps of Man and Selective Effects of Climate'. *Proc. Nat. Acad. Sci. U.S.A.* **78** 2638—2642.

Piazza, A., Rendine, S., Zei, G., Moroni, A., and Cavalli-Sforza, L. L. 1987. 'Migration Rates of Human Populations from Surname Distributions'. *Nature* **329** 714—716.

Pulgram, E. 1958. *The Tongues of Italy*. Cambridge: Harvard University Press.

Rendine, S., Piazza, A., and Cavalli-Sforza, L. L. 1986. 'Simulation and Separation by Principal Components of Multiple Demic Expansions in Europe'. *The American Naturalist* **128** 681—806.

Renfrew, C. 1987. *Archeology and Language. The Puzzle of Indo-European Origins*. Cambridge; Cambridge University Press.

Sherrat, A. 1980. (ed.) *The Cambridge Encyclopedia of Archeology*. Ch. 33, 34. New York: Cambridge University Press.

Wijsman, E. M. and Cavalli-Sforza, L. L. 1984. 'Migration and Genetic Population Structure with Special Reference to Humans'. *Ann. Rev. Ecol. Syst.* **15** 279—301.

Zei, G., Matessi, R. G., Siri, E., Moroni, A., and Cavalli-Sforza, L. L. 1983. 'Surnames in Sardinia. I. Fit of Frequency Distributions for Neutral Alleles and Genetic Population Structure'. *Ann. Hum. Genet.* **47** 329—352.

LUIGI ACCARDI

QUANTUM PROBABILITY AND THE FOUNDATIONS
OF QUANTUM THEORY

1. INTRODUCTION

The point of view advocated, in the last ten years, by quantum probability about the foundations of quantum mechanics, is based on the investigation of the mathematical consequences of a deep and elementary idea developed by the founding fathers of quantum mechanics and accepted nowadays as a truism by most physicists, namely: *one should be careful when applying the rules derived from the experience of macroscopic physics to experiments which are mutually incompatible in the sense of quantum mechanics.*

This statement, like the Dirichlet pidgeonhole principle, looks quite trivial at first sight but some of its consequences are far from evident. It might therefore happen that one accepts the above statement as self-evident but quivers at some other statements which are necessary, if not so evident, corollaries of it.

In order to concretize the general premise above, let us come to the point and sum up, without proof the main conclusions of the quantum probabilistic approach to the foundations of quantum theory.

(1) The basic empirical data, against which we check the predictions of our theories, are conditional probabilities. The conditional probability of an event A given an event C is a model independent notion which can be experimentally estimated by preparing an ensenble of systems for which the event C is realized and measuring on it the relative frequency of A. Since the preparation C preceeds in time the measurement of A, the events C and A can be incompatible in the sense of quantum theory. Given these transition probabilities, all the other relevant quantities: mean values, fluctuations, correlations, . . . , can be obtained by application of standard mathematical formulae.

(2) If $A, B, C, \ldots$ are events and the transition probabilities $P(A \mid B)$, $P(B \mid C)$, $P(C \mid A)$, $\ldots$ are estimated in mutually incompatible experiments, then the unrestricted application of the rules of classical probability to these statistical data is unwarranted. The

R. Cooke and D. Costantini (eds), Statistics in Science. The Foundations of Statistical Methods in Biology, Physics and Economics, 119–147.
© 1990 *Kluwer Academic Publishers.*

existence of a single ·classical probalistic model for these data
(which is equivalent to the possibility of unrestricted application to
these data of the rules of classical probability) cannot be postulated
a priori, but must be proved by means of a mathematical theorem.
The mathematical technique to prove theorems of this kind consists in evaluating the *Statistical Invariants*, which guarantee the
existence (or non existence) of the model uniquely in terms of the
experimental data $P(A \mid B)$, ...

(3) All the existing 'proofs' of the statement that, if a system is in a
superposition state with respect to an observable A, then the
observable A does not actually assume any of its values, are based
on the uncritical application of the rules of the classical probabilistic model to statistical data obtained from mutually incompatible experiments. As such they are wrong. Therefore there is no
rational, nor mathematical, nor experimental, nor theoretical support for the statement 'in a superposition state with respect to the
observable A, the observable A cannot actually assume any of its
values'. Once this statement is rejected, there is no ground to claim
that terms like 'collapse of the wave packet' refer to an actual
change of the physical state of the system.

(4) If one accepts that, in a superposition state for the observable A, A
cannot actually assume any of its values, and that only the act of
measurements 'collapses' the physical state of the system so that
one and only one of the values A is assumed, then one cannot
escape paradoxes of *EPR* type (a measurement here collapses a
particle there) or of the type of the 2-slit experiment (an electron
not looked at is spread thoughout the available space).

(5) Conversely, if one proves that the statement: 'in a superposition
state with respect to A, the observable A cannot actually assume
any of its values' is unwarranted, then all the so called paradoxes of
quantum theory, constructed by means of variations on the theme
'collapse of the wave packet' are swept away.
Quantum probability has provided such a proof.

(6) The fact that quantum mechanics introduced a new kind of probability calculus was recognized since the early days of the theory.
The new elements introduced by quantum probability in this
debate are:
   (i) the individuation of the single axiom of classical probability
        theory whose breakdown, in the quantum domain, can be
        proved uniquely in terms of the experimental data.

(ii) the proof of the fact that a consistent physical interpretation of the new probability calculus can be developed entirely within the conceptual framework of classical physics. There is no need at all to introduce the notion of superposition as a new physical state of matter (unobservable in principle). Contrarily to what stated by Feymann, Heisenberg, Dyson, and many others . . . , no contradiction between theory and experiments can be deduced from the statement that neutrons not looked at pass through one and only one slit, that spins not looked at assume one and only one of their possible values, etc.. . . .

(7) An equivalent formulation of the statement (ii) of point (6) above is: the breakdown of the applicability of classical probability theory to the description of some microscopic phenomena does not imply the necessity of postulating a non classical behaviour of the micro-systems in absence of observations of them. In fact, even for macroscopic systems, the applicability of the classical probabilistic rules cannot be postulated a priori, but should be checked through the comparison of the experimental data with the statistical invariants. The existence of macroscopic physical systems, to which the applicability of the classical laws of probability fails, has been investigated by D. Aerts who has obtained interesting partial results in this direction.

(8) Quantum probability gives a mathematical proof of the fact that no contradiction with the experiments can be deduced from the statement 'electrons not looked at behave exactly as they do when they are looked at' (in particular, they pass through one and only one slit, their spin assumes one and only one value, . . .). With that it satisfies the criterium stated by Heisenberg, Bohr, . . . according to which a good physical theory should not commit itself on situations which are in principle unobservable (e.g. the statement that a neutron not looked at cannot pass through one and only one slit). Any interpretation which accepts that, in a superposition state with respect to the observable A, this observable cannot actually assume any of its values, violates this criterium.

(9) The only conclusion one can draw from the 2-slit experiment is that a set of 3 statistical data (conditional probabilities), obtained in 3 mutually incompatible experiments, cannot be described within a single classical probabilistic model. There is no need to introduce a new 'physical state', in which the position observable has no definite value in order to explain the apparent contradiction arising

from this experiment. Exactly the same conclusion applies to the experiments related to the Bell inequalities. Also in that case one has three different, mutually incompatible experiments which produce statistical data that cannot be described within a single Kolmogorovian model. Bell's statistical data concern correlations and a composite system; the 2-slit experiment concerns conditional probabilities and individual particles. The connection between the two situations escaped Bell's analysis and he was led, like his predecessors and exactly for the same reasons (the implicit postulate that, if the usual physical properties such as locality, objective reality, separability, ... , were true, then the laws of classical probability should be applicable) to belive that a physical property (locality in his case) was responsible of the apparent paradox. It is precisely the validity of this implicit postulate that the quantum probabilistic analysis disproves by means of mathematical and experimental arguments.

One should give Bell the credit of having contributed, with his analysis, to bring back to measurable and observable statements the debate on the foundations of quantum theory, which was sinking into purely formal and mathematical problems. However the limits of his analysis consist in not having recognized that, from the scientific point of view, the famous Bell's inequalities contained nothing new with respect to the old 2-slit experiment.

Quantum probability vindicates the merit of having individuated precisely in this point the root of all the problems concerning the interpretation of quantum mechanics and, more generally, to have realized that no new unplausible physical properties (such as nonlocality, nonreality, ...) need to be introduced to explain the difficulties arising in the interpretation of the experimental data arising in quantum theory, because these are all related to the unwarranted application of a single axiom of probability theory: the Bayes axiom (which in fact, in the usual texts on probability is not even introduced as an axiom, but as the definition of conditional probability). Why this axiom is not a tautology on relative frequencies (as implicitly believed for several centuries) and which deep and far from obvious assumptions it hides, even independently of the experimental results of quantum theory, is explained in [3].

It is the author's hope that the content of this Section clarifies the position of quantum probability within the philosophical debate on the interpretation of quantum theory so to make improbable misunderstandings of this position as those which emerge from the paper [5].

In this paper Van den Berg, Hoekzema and Radder apparently criticize the analysis of the two-slit experiment made in [4]. As a matter of fact, the main point of these authors, namely that one has not the right of postulating the existence of a unique Kolmogorovian model for a set of states coming from a multiplicity of different and mutually incompatible experiments, is precisely the starting point of quantum probability. The large amount of quotations from the best known experts of quantum theory of the past 50 years, insistently reproduced in [1], [2], [3], [4] (and in several other publications and conferences of the author) is aimed precisely at showing that the postulate of the existence of a unique Kolmogorovian model underlying a set of statistical data, regardless of the fact that these data come from a set of compatible experiments or not, is implicit in the analyses of these physicists. The critique against this implicit postulate is precisely the conceptual starting point of quantum probability. In this sense Van den Berg, Hoekzema and Radder attribute to quantum probability the position criticized by it.

The second, conceptually most important, statement of quantum probability, namely that the only existing "proofs" of the "necessity" of introducing the notion of "physical superposition state" are based on this unjustified assumption, is not discussed by the above mentioned authors in [5]. If this statement is false, it should not be difficult to prove it: it is sufficient to find in the literature a different proof.

On the other hand, from the point of view of quantum probability, the error of supporters of the hidden variable theories like Bell and others . . . is to confuse the notion of realism with the notion of "unique Kolmogorovian model": even if we accept the idea of an underlying unique hidden variable space Ω of whose points all the quantum observables are functions (and this does not contradict quantum probability) why should we postulate that on this space there is a unique probability measure, regardless of the preparation conditions?

2. PROBABILISTIC MODELS

When speaking of the probability of an event A we assume that a set of conditions C, under which this probability is evaluated, have been fixed. This probability will be denoted $P(A \mid C)$ and called the 'conditional probability of A given C'. Sometimes it is useful to interpret C as the preparation of an experiment done to verify the occurrence of A. The number $P(A \mid C)$ is a model independent quantity: it has a meaning independently on how it has been evaluated (by means of relative frequencies, bets, opinions, coherence properties, . . .) and independently of the mathematical model one uses to describe the set of events and their probabilities. Sometimes the notion of time ordering will be made explicit by writing $P(A_t \mid C_s)$ where s and t ($s < t$) are real numbers interpreted respectively as time of occurrence of the preparation C and of the event A.

The basic heuristic models, both for events and preparations, that we shall have in mind in the following are given by statements of the type: 'at a given time the values of a given set of observable lie in a given set of numerical intervals'. Thus, for example, the expression $P(A \in I \mid C \in J)$ will denote the probability that the values of the observable A lie in interval I at time t when measured on a system prepared so that at time $s < t$ the observable C lied in the interval J. When J consists of a single point, say $J = c$, the conditional probability

$$P(A \in I \mid C = c)$$

will also be called a *transition probability*. We assume that the probabilities $P(A \in I \mid C \in J)$ have a meaning whenever I and J are arbitrary Borel subsets of the real line. If A is an observable and $I \subseteq \mathbf{R}$ is an interval, the symbol $[A \in I]$ denotes the event that the measurement of A yields a result in I. These events have a natural structure of a Boolean σ-algebra, induced by the Borel σ-algebra of $\mathbf{R}$.

In the following our starting point will be a quadruple of the form:

$$(1) \qquad \{T, \{A(x)\}_{x \in T}, \mathcal{F}_0, \{P(A(x) \in I \mid C)\}\}$$

where
T is a set
$\{A(x): x \in T\}$ is a family of observables
$\mathcal{F}_0$ is a set, called the set of preparation conditions
For each $x \in T$, $C \in \mathcal{F}_0$ and for each interval $I \subseteq R$ the number

$$(2) \qquad P(A(x) \in I \mid C)$$

is the conditional probability of the event $[A(x) \in I]$ given the preparation condition C.

All these data have to be considered as *model independent*, for example they might be experimental data. We are going to describe several different mathematical models for these model independent data and the basic problem we are going to discuss is the following:

Can one discriminate among the different matematical models, describing the data (1), uniquely in terms of the data themselves?

By a mathematical model for the data (1), we mean a correspondence which associates to each event $[A(x) \in I]$ (or observable $A(x)$) and to each preparation condition C a mathematical object in such a way that the model independent probabilities $P(A(x) \in I \mid C)$ are described by some formulas relating among themselves the mathematical objects associated to the events (or the observables) and the preparations. Definition (3) in the following will make this notion precise.

Two such models are called *stochastically equivalent* if they describe the same set of conditional probabilities. Note that this notion of stochastic equivalence is very weak: it only requires that the two models describe the same experimental data — while their mathematical structures can be quite different. For example any set of three 2×2 bistochastic matrices which admits a Kolmogorovian model admits also a complex Hilbert space model (cf. [2]). Hence, for the description of three 2×2 bistochastic matrices, the two models are stochastically equivalent in the sense specified above, even if they are far from being isomorphic as mathematical structures.

Remark (1) When the condition C and the observable $A(x)$ are fixed the map

$$(P(A(x) \in I \mid C) = P_{A(x)}(I \mid C); \qquad I \subseteq \mathbf{R}$$

is a classical probability measure whose numerical characteristics (mean, variance, momenta, characteristic function, etc. . . .) can be defined in the usual way. In particular one defines the mean, or expectation value, of $A(x)$ given C

$$E(A(x) \mid C) = \int_{\mathbf{R}} \lambda \, dP_{A(x)}(\lambda \mid C)$$

the moments of order k of $A(x)$ given C

$$E(A(x)^k \mid C) = \int_R \lambda^k \, dP_{A(x)}(\lambda \mid C)$$

the variance of $A(x)$ given C

$$\mathrm{Var}(A(x) \mid C) = E(A(x)^2 \mid C) - E(A(x) \mid C)^2$$

and the characteristic function of $A(x)$ given C

$$\int_R e^{i\lambda p} \, dP_{A(x)}(\lambda \mid C).$$

Thus, for a single observable (or more generally for a set of compatible observables) and a single conditioning all these models are reduced to the usual one. Only in presence of several observables or several conditionings the nonclassical statistical features arise.

DEFINITION 2. In the above notations, an observable $A(x)(x \in T)$ is called *discrete* if there exists a finite or countable subset $S(x) \subseteq R$ such that, for every $C \in \mathscr{F}_0$ one has

$$(3) \qquad \sum_{a \in S(x)} P(A(x) = a \mid C) = 1.$$

If moreover, for each value $a \in S(x)$, there exists a preparation $C \in \mathscr{F}_0$ such that

$$(4) \qquad P(A(x) = a \mid C) > 0$$

then the set $S(x)$ is called the set of values of the discrete observable $A(x)$.

We shall consider as primitive the notion of observable. Heuristically an observable is defined by a set of measurement procedures and by the set of its values. Usually the values of an observable are real numbers, but sometimes it is convenient to consider a set of (compatible) observables as a single vector valued observable. Two observables $A(x)$, $A(y)$ are called *compatible* if they can be simultaneously measured. This is a model independent, and therefore vaguely defined, notion. In the class of models of von Neumann type (cf. the following Definition 3), this property will correspond to the fact that the algebras $\mathscr{A}(x)$ and $\mathscr{A}(y)$, generated by $A(x)$ and $A(y)$, commute. An observable compatible with any other observable is called *universally compatible* or also a *superselection observable*. An observable $A(x)$ is called a (measurable, continuous, . . .) function of a set

$$\{A(y), \quad y \in F \subseteq T\}$$

of compatible observables if there exists a (measurable, continuous, . . .) function

$$f : \mathbf{R}^F \to \mathbf{R}$$

such that for every measurement of $A(x)$ with result a there is a simultaneous measurement of all the $A(y)$ with result $a(y)$ ($y \in F$) such that $a = f(\{a(y)\}_{y \in F})$. An observable $A(x)$ is called *maximal* if another observable B can be compatible with $A(x)$ if and only if B is expressible as a function of $A(x)$ and of the universally compatible observables (i.e. the reult of each measurement of B can be obtained from a simultaneous measurement of $A(x)$ and of a set of universally compatible observables).

DEFINITION 3. The preparation C is called *pure* with respect to the family of observables $\{A(x) : x \in T\}$ if there exists a discrete maximal observable B and a value b of B such that for all $x \in T$ and all intervals $I \subseteq \mathbf{R}$ one has:

(5) $P(A(x) \in I \mid B = b) = P(A(x) \in I \mid C).$

A preparation pure with respect to all the observables of a given theory is called *pure*.

All the probabilistic models we are going to consider fit into the abstract scheme described by the following definition:

DEFINITION 4. A probabilistic model for the quadruple (1) is defined by a triple:

(6) $\{\mathscr{A}, \{\varphi_C : C \in \mathscr{F}_0\}, \{e_{A(x)}(\cdot)\}\}$

where
$\mathscr{A}$ is a *-algebra over the complex numbers.
For each preparation condition $C \in \mathscr{F}_0$, φ_C is a state on $\mathscr{A}$
For each $x \in T$

(7) $I \subseteq \mathbf{R} \mapsto e_{A(x)}(I) \in \mathscr{A}$

is an $\mathscr{A}$-valued measure on **R** and where, for all x in T, C in $\mathscr{F}_0$ and I $\subseteq$ **R** one has

(8) $P(A(x) \in I \mid C) = \varphi_C(e_{A(x)}(I)).$

The $e_{A(x)}(I)$ correspond, in the mathematical model, to the events $[A(x) \in I]$. If they are orthogonal projections of $\mathscr{A}$, one speaks of a *von Neumann model*; if they are positive elements of $\mathscr{A}$, one speaks of a *Ludwig type model*. In both these type of models the $\mathscr{A}$-valued measure (7) is supposed to be countably additive.

In case of models of von Neumann type, given $x \in T$, the commutative *-algebra generated by the family of all $e_{A(x)}(I)$ with $I \subseteq$ **R** is called the algebra generated by the observable $A(x)$ and denoted $\mathscr{A}(x)$. For probabilistic models not included in Definition 3 above.

As we see in this scheme the roles of the preparations and of the events are quite asymmetric. In several important applications however the conditions of the problem imply a complete symmetry between the notions of preparation and event in the sense that every preparation is a non null event and every non null event is a preparation. To deal with these situations the abstract model discussed above has to be particularized into a finer structure.

3. THE STANDARD QUANTUM MODEL

In the same conditions as in Section (2).

DEFINITION 1. In the class of models defined by Definition (2.3) the *standard* quantum model for the conditional probabilities (2.2) is characterized by:

(1) $\mathscr{A} = \mathscr{B}(H)$

(2) $e_{A(x)}(I) = \chi_I(\bar{A}(x))$

(3) $P(A(x) \in I \mid C) = \varphi_C(e_{A(x)}(I)) = \mathrm{Tr}(W_C \cdot e_{A(x)}(I))$

where

(i) H is a complex separable Hilbert space with scalar product denoted $\langle .,. \rangle$ and antilinear in the first variable and $\mathscr{B}(H)$ denotes the algebra of all the bounded operators on H.

(ii) For each $x \in T$, $\overline{A}(x)$ is a self-adjoint operator on H and

(4) $$\overline{A}(x) = \int_R \lambda \, de_{A(x)}(\lambda)$$

is the spectral decomposition of $\overline{A}(x)$.

(iii) Tr denotes the trace on H, i.e.

(5) $\text{Tr}(X) = \Sigma \langle e_i, Xe_i \rangle;$ (e_i) an o.n. basis of H

W_C is a *density matrix* on H, i.e. a positive trace class operator satisfying $\text{Tr}(W_C) = 1$.

In the sequel we shall use the following equivalent notations:

$$\text{Tr}(W_C \cdot e_{A(x)}(I)) \mid C) = E(\chi_I(A(x)) \mid W_C) = \varphi_C(e_{A(x)}(I))$$

where $\varphi_C(\cdot) = \text{Tr}(W_C \cdot)$.

The operator $\overline{A}(x)$ associated to the observable $A(x)$ is called itself an observable and denoted with the same symbol unless confusion can arise. In the literature on the mathematical foundations of quantum mechanics it is frequently postulated that there exists a one-to-one correspondence between physical observables and self-adjoint operators. In practice the self-adjoint operators to which one can associate a well defined measurement procedure are very few and conversely there are observable quantities (time, temperature, ...) which do not correspond to any self-adjoint operator in the usual (nonrelativistic) quantum models.

Since the triple

$$\{H, \{\overline{A}(x)\}_{x \in T}, W_C\}$$

uniquely determines the model, in the following we shall frequently call such a triple a standard quantum model for the probabilities (3).

DEFINITION 2. A *pure quantum model* for the conditional probabilities (3) is a standard quantum model $\{H, \{\overline{A}(x)\}_{x \in T}, W_C\}$ whose density operator has rank one, i.e. such that there exists a unit vector $\psi_C \in H$ satisfying

$$W_C = \mid \psi_C \rangle \langle \psi_C \mid = P(\psi_C)$$

where $\mid \psi_C \rangle \langle \psi_C \mid = P(\psi_C)$ denotes the rank one projection along the direction of ψ_C i.e., the linear operator on H defined by

(6) $P_\psi : \phi \in \mathscr{H} \to \langle \psi, \phi \rangle \, \psi = P_\psi(\phi) \in \mathscr{H};$

$\psi \in \mathscr{H}, |\psi| = 1.$

Remark 3. We will sometimes use Dirac's notation, according to which a vector ψ in H is denoted by $|\psi >$ (and called a *ket vector*); the corresponding linear functional on $H(\phi \in H \to \langle \psi, \phi \rangle)$ is denoted $\langle \psi |$ (and called a *bra vector*), i.e. the bra vectors are elements of the conjugate Hilbert space $\bar{H}$ of H. In these notations that rank one projection P_ψ defined by (1), is also denoted $|\psi > < \psi|$.

Note that, for a pure conditioning, one has

(7) $P(A(x) \in I \mid C) = \langle \psi_C, e_{A(x)}(I) \cdot \psi_C \rangle = \langle \psi_C, \chi_I(\bar{A}(x)) \cdot \psi_C \rangle.$

Remark 4. In terms of partial Boolean algebras these models can be described as follows:

$\mathscr{F}$ is the partial Boolean algebra of othogonal projectors on the Hilbert space H.

$\mathscr{F}_0$ is the set of density matrices on H, in the case of a standard quantum model, and the set of all rank one projectors on H in the case of a pure quantum model.

All the events considered in Quantum Theory can be reduced to the statement that the results of the measurements of certain observables are numbers lying in certain intervals of the real line, whose width is determined by the precision of the measurement. In the mathematical model an observable is represented by a selfadjoint operator A acting on H and the event that the result of the measure of A is in the interval I by the spectral projector $e_A(I)$. In particular, if I contains a single simple eigenvalue of the operator A, say λ, corresponding to the eigenvector ϕ in the sense that:

(8) $A\phi = \lambda\phi$

then

(9) $e_A(I) = |\phi > < \phi| = P_\phi^A$

If A is an observable satisfying (8), (9) for some interval I, B is any observable and J is any interval of $\mathbf{R}$, then the probability

(10) $\Pr(B \in J \mid A = \lambda) = P(e_B(J) \mid P_\phi^A) = \mathrm{Tr}(e_B(J) \cdot P_\phi^A) =$

$\langle \phi, e_B(J) \cdot \phi \rangle$

is interpreted as the conditional probability that the value of B is in the interval J if the system is prepared so that the value of A is λ. If moreover

$$e_B(J) = P^B_\psi; \qquad B\psi = \mu\psi; \qquad \mu \in \mathbf{R}$$

then the conditional probability

$$(11) \qquad \mathrm{Prob}(B = \mu \mid A = \lambda) = \mathrm{tr}(P^B_\psi \cdot P^A_\phi) = |\langle \psi, \phi \rangle|^2$$

is called a *transition probability*. Notice that, in view of (11), the quantum transition probabilities satisfy the symmetry condition

$$(12) \qquad \mathrm{Prob}(B = \mu \mid A = \lambda) = \mathrm{Prob}(A = \lambda \mid B = \mu)$$

which holds whenever λ and μ are simple discrete eigenvalues of A and B respectively.

Remark 5. When considered as elements of $\mathscr{F}$, rank one projectors on H are called *pure* or *atomic events*; when considered as elements of $\mathscr{F}_0$ they are called *pure preparations* or, more frequently, *pure states*. Rank one projectors lie in the intersection between the preparations and the events and, since they are in one-to-one correspondence with the one-dimensional sub-spaces of H, another way of looking at pure states is as rays in H, i.e. elements of the projective space over H (the set of equivalence classes in $H - \{0\}$ for the relation $\phi \sim \psi \Leftrightarrow \exists \lambda \in \mathbf{C}: \phi = \lambda\psi$).

Often, by abuse of language, one uses the term *state* for a density matrix W_C or, in case $W_C = |\psi_C\rangle \langle\psi_C|$, for the vector ψ_C itself. In the former case one speaks of a *mixed state* or a *mixture*, in the latter of a *pure* or *vector state* or, if H is realized as the complex L^2 space over some measure space, of a *wave function*. The term *pure* refers here to the fact that the state φ_C, corresponding to a rank one density matrix, is extremal in the convex set of all the states on the algebra $\mathscr{B}(H)$. This terminology agrees with the more general one to be introduced in Section (5).

4. TRANSITIONS AMONG DISCRETE OBSERVABLES: THE KOLMOGOROV AND THE RENYI MODELS

The notations and the assumptions are the same as in Section (2). Let S be a finite or countable subset of the natural integers. Assume that, for each $x \in T$ the observable $A(x)$ is discrete, and let $\{a_n\ (x): n \in S\}$ denote its set of values. We choose the set of preparations conditions to coincide with the set of all the events of the form $[A(x) = a_n(x)]$ for some $x \in T$ and $n \in S$. These events will be called *atomic events*.

DEFINITION 1. A *Kolmogorovian model* for the transition probabilities

(1) $P(A(y) = a_n(y)|A(x) = a_m(x)), \ m, n \in S; \ a, y \in T$

is a pair:

(2) $\{(\Omega, \mathscr{F}, \mu), \{A_n(x)\}\ (n \in S)\ (x \in T)\}$

where
(i) $(\Omega, \mathscr{F}, \mu)$ is a probability space.
(ii) For each $x \in T$, $A_n(x)$: $n \in S$ is a measurable partition of Ω and for each $x, y \in T$; $m, n \in S$ one has

(3) $\mu(A_n(x)) > 0$

(4) $P(A(y) = a_n(y)|A(x) = a_m(x)) = \dfrac{\mu(A_n(y) \cap A_m(x))}{\mu(A_m(x))}$

The definition of the Renyi model for the transition probabilities (1) differs from that of Kolmogorov model only in that it is not required that $\mu(\Omega) = 1$ but μ can be an arbitrary positive σ- finite measure. In particular the basic identity (4) holds in both models. In both models the observable $A(x)$ is represented by the function $\bar{A}(x)$: $\Omega \rightarrow \{a_1(x), \ldots, a_n(x), \ldots\}$ taking on $A_n(x)$ the constant value $a_n(x)$.

5. TRANSITIONS AMONG DISCRETE MAXIMAL OBSERVABLES: THE PURE QUANTUM MODEL

For discrete observables the notion of maximality can be introduced in a purely probabilistic way. We denote $P(B \mid A(y) = a_n(y); A(x) = a_m(x))$ the probability of an event B under the condition that at a certain moment the measurement of the observable $A(x)$ has given the result $a_m(x)$ and at a subsequent time the measurement of $A(y)$ has given the result $a_n(y)$ (in both cases the times can be included in the variables x, y and omitted from the relations).

DEFINITION 1. The discrete observables $\{A(x)\}$, $x \in T\}$ are called maximal if for each event B and for all x, $y \in T$, m, $n \in S$ one has:

(1) $P(B \mid A(y) = a_n(y); A(x) = a_m(x)) = P(B \mid A(y) = a_n(y))$.

Condition (1) means that the information contained in the knowledge that the event $[A(x) = a_n(x)]$ has happened can be changed, but not increased, by subsequent measurements. The existence of different sets of maximal observables is the qualitative essence of the *Heisenberg indeterminacy principle*.

According to quantum theory, the transition probabilities (3.11), when referred to the values of a maximal family $\{A(x) \ (x \in T)\}$ are described by the following model:

DEFINITION 2. A *pure quantum model*, or simply a *complex Hilbert space model* in the following, for the transition probabilities (3.11), is a pair:

(2) $\{H, \{\psi_n(x): n \in S \ x \in T\}\}$

where

(i) H is a complex Hilbert space of dimension equal to the cardinality of S (i.e. the number of different values of the $A(x)$).

(ii) for each $x \in T$ the vectors $(\psi_n(x): n \in N)$ form an orthonormal basis of H and for each x, $y \in T$ and m, $n \in S$ one has

(3) $P(A(y) = a_n(y) \mid A(x) = a_m(x)) = |\langle \psi_n(y), \psi_m(x) \rangle|^2$.

If H is a real Hilbert space satisfying (i), then we speak of a *real Hilbert space model*.

Remark 3. Notice that (3) implies that a necessary condition for the family of transition probabilities (3.11) to be described by a quantum model for discrete maximal observables $A(x)$ is that the symmetry condition

(4) $P(A(y) = a_n(y) \mid A(x) = a_m(x)) = P(A(x) =$

$a_m(x) \mid A(y) = a_n(y))$

be satisfied for each x, $y \in T$ and m, $n \in S$.

Remark 4. The maximality of the discrete observable $A(x)$ is expressed, in the quantum model, by the fact that the correspondence between the values $a_n(x)$ of $A(x)$ and the vectors $\psi_n(x)$ of the orthonormal basis is one to one. The statement: 'at time t the system is in the pure state $\psi_n(x)$' means a (idealized, instantaneous) measurement of the observable $A(x)$ at time t has given the result $a_n(x)$.

Notice that in the expression (3) of the probabilities the actual values of the observables $A(x)$ play no role: only the orthonormal basis $(\psi_n(x))$ is used.

In this model one associates to the discrete observable $A(x)$ the self-adjoint operator

$$\overline{A}(x) = \sum_n a_n(x \mid \psi_n(x))\langle \psi_n(x)|.$$

and, due to the identity

$$|\langle \phi, \psi \rangle|^2 = \mathrm{Tr}(|\phi\rangle\langle\phi| \cdot |\phi\rangle)\phi|).$$

The following result shows that a pure quantum model for the transition probabilities $p_{m, n}(x, y)$ can be equivalently defined in terms of *transition amplitudes*.

PROPOSITION 5. *Every pure quantum model* $\{H, \{\psi_n(x): n \in S_{x \in T}\}\}$ *for the transition probabilities* (4) *defines a triple*

(5) $\{H, \{U(x, y)\}_{x, y \in T}, (e_n)\}$

such that

(i) *H is an Hilbert space* (ii) (e_n) *is an orthonormal basis of H* (iii) *For each x, y $\in$ T, U(x, y) is a unitary operator on H satisfying:*

(6) $U(x, y)^{-1} = U(x, y)^* = U(y, x); \qquad U(x, x) = 1$

(7) $U(y, z) U(x, y) = U(x, z) \qquad z \in T$

(iv) *For each x, y$\in$ T and m, n$\in$ S one has:*

(8) $P(A(y) = a_n(y) \mid A(x) = a_m(x)) = |\langle e_n, U(x, y) e_m \rangle|^2 = $

        $|U_{n, m}(x, y)|^2$

(*where, by definition,* $\langle e_n, U(x, y)e_m \rangle = U_{n, m}(x, y)$. *Conversely any triple* (5) *defines a pure quantum model for the transition probabilities* (4) *via the correspondence*

(9) $\psi_n(x) = U(x_0, y)e_n, \quad n \in S, \ x_0 \in T\text{-fixed arbitrarily.}$

Proof. For each x, $y \in T$ denote $U(x, y)$; $H \rightarrow H$ the unitary operator characterized by

(10) $U(x, y)\psi_n(x) = \psi_n(y)$; $n \in N$.

Then, by definition for any x, $y \in T$ (6) and (7) hold. Moreover, if $x_0 \in T$ is an arbitrary fixed element of T, then denoting

(11) $e_n = \psi_n(x_0)$

and using the unitarity of $U(x_0, y)$ and (7) one finds

(12) $|\langle e_n, U(x, y)e_m \rangle|^2 = |\langle U(x_0, y)e_n, U(x_0, y)U(y, x)e_m \rangle|^2$

$= |\langle \psi_n(y), U(x_0, x)e_m \rangle|^2 = |\langle \psi_n(y), \psi_m(x) \rangle|^2$.

Thus (8) follows from (3). It is clear from the construction that the triple (5), with $U(x, y)$ defined by (10) and e_n by (11), satisfies (6), (7), (8) and uniquely determines the quantum model

$$\{H; \psi_n(x); \quad n \in S, \quad x \in T\}.$$

Conversely, given any such a triple one can define, for each $x \in T$, the orthonormal basis

$$\psi_n(x) = U(x_0, x)e_j$$

where $x_0 \in T$ is fixed once for all and independent on T. Since clearly (3) holds by assumption it follows that the resulting pair is a pure quantum model for the transition probabilities (4) in the sense of Definition (2).

DEFINITION 6. A family $\{U(x, y)_{x, y \in T}\}$ of unitary operators on a Hilbert space H satisfying conditions (6), (7) is called a family of *transition amplitudes* on T. The matrix elements

(13) $U^\psi_{n, m}(x, y) = \langle e_n, U(x, y)e_m \rangle = \langle \psi_n(x), \psi_m(y) \rangle$

are called the *transition amplitudes between the states* $\psi_n(x)$ and $\psi_n(y)$.

Remark 7. Note that (7) implies that

(14) $U^\psi_{n, m}(x, z) = \sum_k U^\psi_{n, k}(x, y)U^\psi_{k, m}(y, z)$

which, by analogy with the classical theorem of composite (or total) probabilities will be called *the theorem of composite (or total) ampli-*

tudes. Taking the square modulus of both sides of (14) with the simplifying notation

$$(15) \quad P(A(y) = a_n(y) \mid A(x) = a_m(x)) = p_{n,m}(y, x) = p_{m,n}(x, y)$$
$$= |U_{n,m}(x, y)|^2$$

one finds

$$(16) \quad p_{n,m}(x, z) = \sum_k p_{n,k}(x, y) p_{k,m}(y, z) +$$

$$+ \sum_{k \neq k'} \bar{U}^\psi_{n,k}(x, y) \bar{U}^\psi_{n,k'}(x, y) U^\psi_{m,k'}(y, z).$$

On the other hand, repeating the construction of Remark (5.3) with the identifications:

$$A = |\psi_n(y)\rangle\langle\psi_n(y)|; \qquad \psi_j = \psi_j(z); \qquad \phi = \psi_m(x)$$

one finds:

$$(17) \quad E(A \mid \phi) = E(A \mid W_\phi) = p_{n,m}(x, z) - \sum_k p_{n,k}(x, y) p_{k,m}(y, z)$$

$$= \sum_{k \neq k'} \bar{U}^\psi_{n,k}(x, y) \bar{U}^\psi_{k,m}(y, z) U^\psi_{n,k'}(x, y) U^\psi_{m,k'}(y, z).$$

The terms (17) are called the *interference terms.*

Since in a classical model satisfying (1), one should have:

$$\sum_k p_{n,k}(x, y) p_{k,m}(y, z) =$$

$$\sum_k P(A(x) = a_n(x) \mid A(z) \qquad = a_k(z)) \cdot P(A(z) = a_k(z) \mid A(y) = a_m(y)) =$$

$$\sum_k P(A(x) = a_n(x) \mid A(z) \qquad = a_k(z); A(y) = a_m(y)) \cdot P(A(z) = a_k(z) = a_m(y)) =$$

$$P(A(x) = a_n(x) \mid A(y) = a_m(y))$$

it follows that:

(i) in the quantum model for the transition probabilities among discrete maximal observables the theorem of composite probabilities does not hold.

(ii) the interference terms (17) are a quantitative measure of the deviation from the validity of the theorem of the composite probabilities. Since both formulas (1) and (8) can be experimentally verified, they provide a simple experimental test for the existence of a Kolmogorovian model for the transition probabilities (4).

6. TRANSITION PROBABILITIES, EVOLUTIONS, ENERGY

In this Section a purely probabilistic approach to the notion of 'energy' in quantum theory is given and it is shown that, even in the most elementary quantum mechanical context in which this notion appears there are some physically unjustified assumptions in the way this notion is introduced. The alternative mathematical models of quantum theory which do not introduce this assumption are worth investigation both from the mathematical and the physical point of view.

In the notations of the previous Section, let us now choose T to be the real line whose elements we interpret as time. In this case the observable $A(t)$ $(t \in T)$ can be interpreted as the time t translate of a given fixed observable A. We can assume that the set of possible values of $A(t)$ does not depend on t. Let us denote $(a_n)_{n \in S}$ this set of values. In the theory of classical Markov chains the transition probability matrices $P(s, t)$ $(s \leqslant t)$ satisfy the evolution equation

(1) $P(r, t) = P(s, t) P(r, s)$ $r < s < t$

called the Chapman-Kolmogorov equation. The analogue of this evolution equation for the transition amplitudes is the Schrödinger equation.

Let $\{H, \{U(s, t)\}_{s, t \in R}, (e_n)\}$ be a pure quantum model for the transition probabilities (4) in the sense of Proposition (5.5), then the Equations (6), (7) imply that

(2) $U(s, t) = U(t, s)^{-1};$ $U(t, t) = 1$

(3) $U(s, t) U(r, s) = U(r, t)$ $r < s < t.$

Because of (2) it is sufficient to restrict one's attention to the $U(s, t)$ with $s < t$.

DEFINITION 1. A 2-parameter family $U(s, t)$ $(s \leqslant t)$ of unitary operators satisfying (2) and (3) will be called a *left unitary evolution*. If condition (3) is replaced by

(4) $U(r, s) U(s, t) = U(r, t)$ $r < s < t$

then we speak of a *right unitary evolution*. For future reference recall that, in probabilistic language, a family of random variables $U(s, t)$, satisfying (13) (resp. (14)), is called a left (resp. right) multiplicative functional on **R**. Unless stated otherwise, by unitary evolution we shall mean right unitary evolution.

LEMMA 2. *Let D be a dense subspace of H such that for each $\xi \in D$ and for any $n \in$ **R** the map $t \in [s, +\infty] \mapsto U(s, t) \xi$ is strongly differentiable with derivative $\partial_t U(s, t)\xi$. Then:*

  (i) *For each $s < t$ there exists an operator $H_s(t)$ symmetric on $U(s, t) D$ such that*

(5) $\partial_t U(s, t)\xi = iH_s(t)U(s, t)\xi$

  (ii) *For each $r < s < t$ one has*

(6) $H_r(t) = U(z, s)H_s(t)U(r, s)^*$

on $U(r, t) D$.

Proof i. follows from (3) and differentiation of the (constant) map

$$t \mapsto \frac{\delta}{\delta t} \langle U(s, t)\xi, U(s, t)\eta \rangle, \quad \xi, \eta \in D.$$

Since the $U(s, t)$ are unitary operators, the existence of $D_t U(s, t)\xi$ in the strong sense implies the existence of $D_t U(r, t)\xi$ for any $r \leqslant s \leqslant t$ and the identity

$$\partial_t U(r, t)\xi = \partial_t U(s, t) \cdot U(s, t)\xi$$

hence, for each $\eta \in U(r, t)D$

$$H_r(t)\eta = U(r, s)H_s(t)U(r, s)^*\eta$$

and this proves (ii).

We shall refer to the result of Lemma (2) above by saying that the right evolution $U(s, t)$ satisfies on D the equation

(7) $\dfrac{d}{dt} U(s, t) = iH_s(t)U(s, t);$ $U(s, s) = 1.$

Equation (7) is called the *Schrödinger equation with time dependent hamiltonian* $H_s(t)$ or, according to the interpretation of $U(s, t)$, the *Schrödinger equation in the interaction representation.*

The condition of stationarity of the transition probabilities

(8) $P(A(t + r) = a_n \mid A(s + r) = a_m) = P(A(t) = a_n \mid A(s)$

        $= a_m)$

$(s < t, r \in R, m, n \in S)$ becomes in view of (8):

(9) $|\langle e_n, U^A(s + r, t + r), e_m \rangle|^2 = |\langle e_n, U^A(s, t)e_m \rangle|^2$

where $(U^A(s, t))$ is the transition amplitude associated to the pair $\{H, (\psi_n(A, t))\}$ according to Proposition (5.5). It is a difficult problem to find the conditions under which the validity of (9), for all $s < t$, $r \in R$, m, $n \in S$ and for 'sufficiently many' orthonormal bases (e_n), implies that, up to a phase

(10) $U^A(s + r, t + r) = U^A(s, t).$

Assuming however the validity of (9) for all $s < t$ and $r \in R$, and defining

(11) $U_t^A = U^A(0, t)$

(3) implies that one has that (U_t^A) is a 1-parameter unitary group on H. Thus, if (U_t^A) is strongly continuous, then there exists a self-adjoint operator h^A on H such that

(12) $\dfrac{\mathrm{d}}{\mathrm{d}t} U_t^A = ih^A U_t^A$

conversely, via Stone's theorem, any self-adjoint operator h^A on H defines a 1-parameter unitary group U_t^A on H satisfying the Equation (12).

Remark 3. Notice that the transition amplitudes $U^A(s, t)$, associated to the pure quantum model for the transition probabilities of the observables $A(t)$ depend, *a priori*, on the observable A and its time translate. If we have two pure quantum models

(13) $\{H, \{\psi_n(A, t); \ n \in S, \ t \in R\};$ $\{H, \{\psi_n(B, t); \ n \in S,$
          $t \in R\}$

relative to two discrete maximal observables A, B and their time translates, then the corresponding transition amplitudes, defined by

$$(14) \qquad U^x(x, t)\psi_n(X, s) = \psi_n(X, t)$$

$(X = A, B, s < t, n \in S)$ are related by

$$(15) \qquad V_{AB}(t)^{-1} U^B(s, t) V_{AB}(s) = U^A(s, t)$$

where $V_{AB}(t)$ is the unitary operator on H defined by:

$$(16) \qquad V_{AB}(t)\psi_n(A, t) = \psi_n(B, t) \qquad n \in S.$$

One of the basic, and not always explicitly stated, postulates of quantum theory is that the transition amplitude $U^A(s, t)$ is in fact independent on the discrete maximal observables A. Maybe it is possible, in a relativistic and field theoretical context, to deduce the evolution postulate below from considerations of invariance with respect to a given symmetry group, but up to now this problem does not seem to have been investigated.

EVOLUTION POSTULATE. Denote $\mathcal{O}_0$ the family of all (discrete) maximal observables at time O and let $T = \mathcal{O}_0 \times R$. The pair $(A_0, t) \in T$ will be denoted A_t. Then

(P1) For each t the set of values of A_t coincides with the set of values of a_0. The common index set for the values of each A_t will be denoted S.

(P2) Given a pure quantum model $\{H, \{\psi_n(A, t), (n \in S, (A, t) \in T\}\}$ there exists a unitary evolution $U(s, t)$ on H such that for any $s < t$, any observable $A \in \mathcal{O}_0$ and any observable B-not necessarily discrete — with spectral measure E^B (.), the conditional probability that $[B_t \in I]$ $(I \subseteq R$ — a Borel set) given that $A_s = a_n$ is given by:

$$(17) \qquad P(B_t \in I \mid A_s = a_n) = \langle U(s, t)^*\psi_n(A, s),$$
$$E^B (I)U(s, t)^*\psi_n(A, s) \rangle.$$

Notice that, in case also the observable B is discrete and the set I is reduced to a single point b_n, the identity (17) becomes

$$(18) \qquad P(B_t = b_n \mid A_s = a_n) = |\langle U(s, t)\psi_n(B, s), \psi_n(A, s)\rangle|^2.$$

It is instructive, to understand the role of the evolution postulate, to compare this formula with (8) and (12). Recalling that in our case $T =$

$\mathscr{O}_0 \times \mathbf{R}$ according to (8) one should have

(19) $P(B, t) = (b_n, t) | (A, s) = (a_m, s)) =$

$|\langle \psi_n(A, s), U((A, s), (B, t)) \psi_m(A, s) \rangle|^2$

where we have put in (30) $x_0 = (A, s)$. Thus the basic content of the Evolution Postulate in quantum theory is that, independently on the observables A, B one has

(20) $U((A, s), (B, t)) = U(s, t)$.

It is easy to build examples in which (20) is not satisfied. It is however not known if, or under which additional conditions, these constraints imply the identity (20).

DEFINITION 4. Assuming the evolution postulate and the time-homogeneity condition

(21) $U(s + r, t + r) = U(s, t)$.

Then 1-parameter family

(22) $U_t = U(0, t)$

is a strongly continuous unitary group whose generator H is called the *Hamiltonian* of the system and is a self-adjoint operator characterized by the property:

(23) $U_t = \exp it H$.

Remark 5. The probabilistic meaning of the Hamiltonian is illustrated by the following considerations: according to (18) and under the time homogeneity condition (21), the probability that a discrete observable A at time zero takes the value a_m and at time t the value a_n, is given by:

(24) $P(A_t = a_n \mid A_0 = a_m) = |\langle \psi_n, U_t \psi_m \rangle|^2$

where we have put

(25) $\psi_n = \psi_n(A, 0)$.

Assuming that for each m the function $t \mapsto U_t \psi_m$ is twice continuously weakly differentiable and expanding the right hand side of (22) up to the second order included, one obtains, if $m \neq n$

(26a) $P(A_t = a_n \mid A_0 = a_m) = t^2 |\langle \psi_n, H \psi_m \rangle|^2 + O_{m, n}(t^3)$

and if $m = n$

(26b) $P(A_t = a_m \mid A_0 = a_m) = 1 - t^2 \operatorname{Var}(H \mid \psi_m) + O_m(t^3)$

where H is the Hamiltonian of the system and where, $\operatorname{var}(H \mid \psi_m)$ is defined in Section 2.

From (24) it is clear that the operator H is a measure of the rate of transitions between the values a_m and a_n of the observable A (one often says 'between the states ψ_m and ψ_n'). However it is not a rate in the usual sense, since it varies like t^2 and not like t. The operator H represents an important physical quantity called the *energy* of the system. Now let B be any observable and let U_t be given by (23). The self-adjoint operator

(27) $B(t) = U_t \cdot B \cdot U_t^*$

is called the *Heisenberg evolved of B at time t*. The differential form of equation (27) i.e.

(28) $\dfrac{d}{dt} B(t) = i[H, B(t)]; \qquad B(0) = B$

is called the *Heisenberg equation of motion*. According to the evolution postulate in the form (17) and to the spectral theorem, and using the general terminology introduced in Remark (2.1), if f is any Borel function on **R** the conditional mean value of the observalbe $f(B)$ at time t, given that the observable A has the value a_m at time zero (or simply the mean value of $B(t)$ in the state ψ_m), is given by

(29) $E(f(B(t)) \mid A(0) = a_m) = \langle \psi_m, f(B(t)) \cdot \psi_m \rangle$

where ψ_m is the eigenvector of $A = A(0)$ corresponding to the eigenvalue a_m. In particular, if $B = H$, then from (26), (27) and (29), it follows that, for any $t \triangleright \mathbf{R}$, for any observable A and any of its values a_m:

(30) $E(f(H(t)) \mid A(0) = a_m) = E(f(H(0)) \mid A(0) = a_m)$

$= \text{constant}$

which is the quantum mechanical form of the *conservation of the energy*.

From (26b) it follows that for short times t, the smaller the conditional variance of the energy given that $A = a_m$, the higher the probability that the observable A will keep the value a_m after t seconds. By the Schwartz inequality

$$(31) \quad |E(H \mid \psi_m)| = |\langle \psi_m, H\psi_m \rangle| \leqslant \| \psi_m \| \cdot |H\psi_m\|$$

$$= (E(H^2 \mid \psi_m))^{\frac{1}{2}}$$

hence the minimum value of the conditional variance is zero and the equality in (31) (corresponding to the minimum variance) holds if and only if ψ_m and $H\psi_m$ are proportional, i.e. for some scalar E_m (necessarily real and equal to $E(H \mid \psi_m)$) one has:

$$(32) \quad H\psi_m = E_m \psi_m.$$

From (17) and (26) it follows that if ψ_m satisfies (16) then for each n, for each $t \in \mathbf{R}$, for each observable B, and for each Borel function f on $\mathbf{R}$ one has

$$(33) \quad E(f(B(t)) \mid H(0) = E_m) = E(f(B(0)) \mid E_m)$$

$$= \langle \psi_m, f(B(t)) \cdot \psi_m \rangle = \text{constant}$$

which is the dual formulation of the conservation of energy (30). For this reason a value E_m, satisfying (16) is also called a *stationary value* and the corresponding ψ_m a *stationary state* of the system.

7. LOCALITY AND BELL'S INEQUALITY

In this Section we correct some minor algebraic errors contained in Corollary (2) and Theorem (3) of [1] (The identities (4) and (7) of this theorem are equivalent to (2) and (3) and respectively (5) and (6) only in the case of observables taking only the values ± 1. In the general case they are only a consequence of the other two. Moreover the discussion of the beads example in that paper is made obscure by quite a number of typing misprints, therefore we reproduce here a corrected version of this discussion).

Let us recall the following Lemma from [1].

LEMMA 1. *Let a, b, c, be numbers in the interval* $[-1, +1]$. *Then*

$$(1) \quad ab - bc + ac \leqslant 1.$$

COROLLARY 2. *For any given a, b, c, d in the interval* $[-1, 1]$, *the following two equivalent inequalities hold*:

(2) $|ab - bc| \leqslant 1 - ac$

(3) $|ab + bc| \leqslant 1 + ac$

and imply

(4) $|ab + bc| + |ad - dc| \leqslant 2.$

If a, b, c, d are restricted to the set $\{+1, -1\}$ then (2), (3) and (4) are equivalent.

Proof. If $ab - bc = 0$, then (2) follows form Lemma (1). If $ab - bc < 0$, then (2) is equivalent to

$$bc - ab + ac \leqslant 1$$

and, with the substitutions

$$a' = c; \qquad b' = a; \qquad c' = b$$

the above inequality becomes

$$a'b' - b'c' + a'c' \leqslant 1$$

which again follows from lemma (1). Thus (2) is true. Since (3) is obtained from (2) exchanging c into $-c$, is follows that (3) is equivalent to (2). Adding (3) to (2) gives (4). Finally, if a, b, c, d are restricted to the set $\{+1, -1\}$ then, letting in (4) $d = a$ one finds (2) and this proves the equivalence.

THEOREM 3. Let (Ω, F, P) be a probability space, and let A, B, C, D be any four random variables defined on Ω and taking values in the interval $[-1, 1]$. Then the following equivalent inequalities hold:

(5) $|E(AB) - E(BC)| \leqslant 1 - E(AC)$

(6) $|E(AB) + E(BC)| \leqslant 1 + E(AC)$

and imply

(7) $|E(AB - BC)| + |E(AD - DC)| \leqslant 2$

where E denotes expectation with respect to the P-measures, i.e.

$$E(F) = \int_{\Omega} F \, dP; \qquad F \in L^{\infty}(\Omega, \mathscr{F})$$

If A, B, C, D take values in the set $\{+1, -1\}$ then the three inequalities are equivalent.

Proof. Corollary (2) implies that

(8) $|AB - BC| \leqslant 1 - AC;$ $|AB + BC| \leqslant 1 + AC;$

$|AB - BC| + |AD + DC| \leqslant 2.$

Thus (5), (6) and (7) follow by taking expectations of both sides of the inequalities (8) and using the inequality $|E(F)| \leqslant E(|F|)$. To prove their equivalence one argues as in the proof of Corollary (2).

The following example, which is a variant of one due to E. Nelson [6], shows that improper application of statistical locality may lead to contradictions independently on the assumption of an unique underlying Kolmogorovian model. It is interesting that Nelson has been the first author to produce an inequality of Bell type not subject to the critique of the implicit assumption of the uniqueness of the Kolmogorovian model.

Consider a set of ± 1 observables S_x with $x = a$, b, c and the associated joint probabilities

(9) $P_{x,y}(S_x^1 = 1; S_y^2 = 1);$ $x, y = a, b, c$

Thus here neither a common preparation nor a single Kolmogorovian model is assumed. Suppose that these probabilities satisfy the *singlet condition*:

(10) $P_{x,x}(S_x^1 = +1; S_x^2 = -1) + P_{x,x}(S_x^1 = -1; S_x^2 = +1) = 1$

and, for each $x \in S^{(2)} \in \mathbf{R}^3$, denote

(11) $P_x = P_{x,x}(S_x^1 = +1).$

LEMMA 4. *Let $x \in S^{(2)}$ and suppose that the probabilities* (9) *satisfy the singlet condition* (10). *Then the following two statements are incompatible*:

(i) *There exists a neighborhood U of x such that the maps*

(12) $x \in U \mapsto P_x$

(13) $y \in U \mapsto P_{x,y}(S_x^1 = i; S_y^2 = j)$

$(i, j = 1$ $y \in S^{(2)})$ *are continuous.*

(ii) *There exists a neighborhood of V of x such that for each $y \in V$,* *the statistical locality condition*

(14)　　$P_{x,y}(S_x^1 = +1; S_y^2 = -1) = P_{x,x}(S_x^1 = +1) \cdot P_{y,y}(S_y^2 = -1)$

holds.

Proof. First notice that (10) is equivalent to

$$O = P_{x,x}(S_x^1 = +1; S_x^2 = +1) = P_{x,x}(S_x^1 = -1; S_x^2 = -1)$$

so that

(15)　　$P_{x,x}(S_x^1 = +1) = P_{x,x}(S_x^1 = +1; S_x^2 = -1) = P_{x,x}(S_x^2 = -1).$

Applying statistical locality)14) and (11), we find

(16)　　$P_{x,y}(S_x^1 = +1; S_y^2 = -1) + P_{x,y}(S_x^1 = -1; S_y^2 = +1) =$

$P_x(1 - P_y) + (1 - P_x)P_y.$

By continuity and the singlet condition, the left hand side of (16) tends to 1, while the right hand side tends to $2P_x(1 - P_x)$ which is less than 1/2.

Remark. The above result is intuitively obvious: (10) means total correlation between the observables S_x^1 and S_x^2, while (14) means total absence of correlations between S_x^1 and S_y^2 for every (x, y), no matter how near y is to x. It is clear that by postulating both relations one introduces a discontinuity in the theory. Therefore, given this conservation law, it is physically unreasonable to postulate the statistical locality property (14). The experiments of Aspect, Rapisarda, . . . show that in some cases this postulate is not only physically unreasonable, but also experimentally false.

ACKNOWLEDGMENTS

The author wants to thank Rudolph Kalman for several remarks on the topics dealt with in the present paper. He also acknowledges support from Grants AFOSR 870249, ONR N00014-86-K-0538, DAAL03-89-0018 through the Center for Mathematical System Theory, University of Florida.

Centro Matematico v. Valterra
Universitá di Roma II, Roma, Italy

BIBLIOGRAPHY

[1] Accardi, L. 1987. 'Foundations of Quantum Mechanics: A Quantum Probabilistic Approach'. In *Quantum Paradoxes*, ed by G. Tarozzi and A. van der Merwe. Dordrecht: Kluwer Acad. Publ.

[2] Accardi, L. and Fedullo, A. 1982. 'On the Statistical Meaning of Complex Numbers in Quantum Theory: *Lettere al Nuovo Cimento* **34** 161–172.

[3] Accardi, L. 1989. 'The Axioms of Probability Theory'. Conference given at the Erice school on Statistics and Probability. E. Regazzini (ed.).

[4] Accardi L. (1984). 'The Probabilistic Roots of the Quantum Mechanical Paradoxes'. Diner et al. (eds) Dordrecht, Kluwer Acad. Publ.

[5] van den Berg H., Hoekzema D., Radder H. 'Accardi on Quantum Theory and the "fifth axiom" of Probability'. *Philosophy of Science* **57** (1990) 149–157.

[6] Nelson E. 'The Locality Problem in Quantum Mechanics' In: *New techniques and ideas Quantum Measurement Theory*. Preprint January 1986.

ALEXANDER BACH

INDISTINGUISHABILITY, INTERCHANGEABILITY, AND INDETERMINISM

1. INTRODUCTION

Before the formalism of quantum theory was introduced, Bose and Einstein in 1924 and, without using this formalism, Fermi in 1925 introduced statistics for indistinguishable particles. Due to the fact, however, that indistinguishable particles were immediately incorporated into the new theory by Dirac in 1926, indistinguishability has always been considered as a typical quantum phenomenon which transcends classical conceptions.

As for the description of the particles of Bose—Einstein (BE) and Fermi—Dirac (FD) statistics the formalism of quantum theory is not needed it is evident that the system which is described by these classical probability distributions is a system of classical particles. But does a concept of indistinguishable classical particles make sense? According to a familiar argument, classical particles, even if they are identical, can always be distinguished by their trajectories. From this viewpoint indistinguishable classical particles do not exist.

Indistinguishable quantum particles are characterized by the the property that the statistical operator which determines the state of the system is invariant under any permutation of the particles. A classical analog of this criterion seems not to exist as traditionally BE and FD statistics are defined as uniform probability distributions on the (different) sets of occupation numbers of these particles. This implies that a description of the individual objects, which is needed for the formulation of invariance under permutations, is not given.

This apparent paradox situation even becomes more confused when Maxwell—Boltzmann (MB) statistics is considered. The particles of MB statistics traditionally are considered as distinguishable such that a description of the individual objects exists. Considering the situation where n particles are distributed on d cells MB statistics is defined as the uniform distribution on the d^n different configurations. Applying now the classical analog of the quantum definition of indistinguishability it turns out that the particles of MB statistics are indistinguishable as any uniform distribution is invariant under permutations.

R. Cooke and D. Costantini (eds), Statistics in Science. The Foundations of Statistical Methods in Biology, Physics and Economics, 149—166.
© 1990 *Kluwer Academic Publishers.*

A solution of these problems is proposed in this lecture from two viewpoints. First, considering permutation invariance of the state as the ultimate criterion for indistinguishability it is shown that BE and FD statistics can be defined on the level of the d^n different configurations by means of a symmetric probability distribution from which the distribution of the occupation numbers is derived. Using instead of symmetric probability measures the equivalent theory of interchangeable random variables, integral representations of states by means of various extensions of de Finetti's theorem are derived.

Second, a careful historical analysis reveals that BE statistics was introduced nearly fifty years before Bose. In his famous memoir of 1877 Boltzmann established the connection between probability and entropy by means of BE statistics. Whereas Boltzmann's entropy concept is equivalent to Shannon's entropy of a probability distribution, the entropy concept of old quantum statistics, $S = k \ln W$, where W is considered as the number of different distributions of n particles on d cells' is the origin of that concept of indistinguishability which is based on an ontological identity of different events and denies the possibility of a description on the level of configurations.

It is a pleasure for me to thank the organizers of this conference, and in particular Professor D. Costantini, for their invitation and the kind hospitality at Luino.

2. INDISTINGUISHABLE QUANTUM PARTICLES

In this section we recall briefly some basic facts of the quantum theory of indistinguishable particles (cf. e.g. [1]).

DEFINITION 2.1. A system of n particles described by the Hilbert space $\otimes^n H$ and the Hamiltonian $h(A_1, \ldots, A_n)$ where A_i, $1 \leq i \leq n$, are copies of a set A of canonical observables of the one-particle system is called a system of *identical particles* if h is invariant under any permutation of the tuples A_i, $1 \leq i \leq n$.

In the following S_n denotes the group of permutations of the integers $\{1, \ldots, n\}$ and for any $\pi \in S_n$ we denote by U_π the familiar unitary permutation operators acting on $\otimes^n H$.

DEFINITION 2.2. A system of n particles is called a system of *indistinguishable particles* if (i) the particles are identical and if (ii) the system is in a state which is characterized by a statistical operator W which is symmetric, i.e.

$$(1) \qquad U_\pi \, WU_\pi^+ = W$$

holds for all $\pi \in S_n$.

Remarks. (1) A property is called extrinsic (intrinsic) if it depends (depends not) upon the state of the system. Accordingly, identical particles agree in all intrinsic properties such as mass, charge etc. (2) According to our definition it is necessary to distinguish between identity and indistinguishability. Identical particles which can be localized in non overlaping regions of space are distinguishable. In particular there is no principle of indistinguishability of identical particles. In the canonical thermal equilibrium state, however, identity implies indistinguishability. (3) The fact that indistinguishable particles are labeled by a number is a property of the description and not an empirical property. (4) We stress that the permutation operators U_π have a passive meaning only such that the action of U_π on a state vector does not describe a physical process.

DEFINITION 2.3. Indistinguishable quantum particles are called (i) bosons if the statistical operator W is *Bose—Einstein symmetric*

$$(2) \qquad U_\pi \, W = W \qquad \text{for all} \qquad \pi \in S_n,$$

(ii) fermions if the statistical operator W is *Fermi—Dirac symmetric*

$$(3) \qquad U_\pi \, W = \text{sgn}(\pi) \, W \qquad \text{for all} \qquad \pi \in S_n.$$

Nonrelativistic quantum statistics is based on the following principle.

SYMMETRIZATION POSTULATE. Indistinguishable particles are either bosons or fermions.

Remarks. (1) A pure state described by the state vector $\Psi \in \otimes^n H$ is symmetric if Ψ is either symmetric (then the state is BE-symmetric) or antisymmetric (then the state is FD-symmetric). Equations (2) and (3) characterize convex combinations of permutation invariant pure states which are built up either from symmetric or antisymmetric vectors. (2) A statistical operator which satisfies (2), (3) respectively, commutes with the orthogonal projection onto the BE (FD) symmetric subspace of $\otimes^n H$ such that the definition of this operator can be restricted to this subspace. (3) In the abstract setting proposed so far a proof of the symmetrization postulate is impossible. Such a proof requires some more structure. i.e. the specification of the configuration space. For $\otimes^n H = \otimes^n L^2(\mathbb{R}^f)$ the symmetrization postulate for pure states is a theorem for $f \geq 3$ but does not hold for $f < 3$. For more information we refer to [2, 3]. (4) Whereas fermions are always statistically dependent there exists a class of statistical independent bosons which is characterized by statistical operators that are homogeneous product states of pure states, i.e.

(4) $W = \otimes^n w$

where $w = w^2$ is a statistical operator on H.

DEFINITION 2.4. A statistical operator of a system of indistinguishable particles is called *Maxwell–Boltzmann symmetric* if W is a homogeneous product state.

Remarks. (1) The statistical operator of independent bosons (4) is MB symmetric. (2) A mixed MB symmetric state is neither FD nor BE symmetric. An example for such a state is the thermodynamical equilibrium state of the strong coupling BCS model [4]. (3) The grand canonical statistical operator for quantum MB statistics is defined on Fock space $F(H) = \oplus_{n=0}^\infty \otimes^n H$

(5) $W_{gc} = Z_{gc}^{-1} \sum_{k=0}^{\infty_\oplus} \frac{1}{k!} \otimes^k \exp(-\beta h + \beta \mu)$

where h is the one-particle hamiltonian. In the high-temperature limit the grand canonical partition function for fermions or bosons converges to Z_{gc}.

3. INDISTINGUISHABLE CLASSICAL PARTICLES

In analogy to the procedure in quantum theory indistinguishable classical particles can be defined as identical classical particles which are in a state, characterized by a probability measure on phase space, which is invariant under permutations (symmetric).

This definition is convenient in a hamiltonian setting and is used, e.g. in [3, 5], for the quantization of indistinguishable particles. As we are concerned with a discrete setting where no hamiltonian is defined we prefer another, in fact equivalent, approach (cf. [6]).

DEFINITION 3.1. Random variables X_i, $1 \leq i \leq n$, defined on a probability space (Ω, F, P) are called *interchangeable* (exchangeable) if

(6) $P(X_i < x_i, 1 \leq i \leq n) = P(X_{\pi(i)} < x_i, 1 \leq i \leq n)$

holds for any $\mathbf{x} = (x_1, \ldots, x_n) \in \mathbb{R}^n$ and any $\pi \in S^n$. A countably infinite sequence of random variables is called interchangeable if (6) holds locally (i.e. for any finite subset).

Remarks. (1) Interchangeable random variables are identically distributed and describe identical objects or events. These objects or events are indistinguishable if, in addition, the probability distribution is invariant under permutations. (2) Random variables are interchangeable if and only if the image measure of P under the mapping $\mathbf{X}$ defined on $\mathbb{R}^n$ is symmetric. (3) Independent identically distributed random variables are interchangeable. (4) For a review of the concept of interchangeability (without any applications to physics) cf. e.g. [7].

In the following we are concerned with the familiar statistical scheme where n particles are distributed on d cells. For the description we introduce a probability space (Ω, F, P) and random variables $X_i: \Omega \rightarrow \{1, \ldots, d\}$, $1 \leq i \leq n$, where $[X_i = j]$ denotes the event that particle i is in cell j. There are alltogether d^n different events $[\mathbf{X} = \mathbf{j}]$, $\mathbf{j} \in \{1, \ldots, d\}^n$, which are called configurations.

DEFINITION 3.2. The particles of the statistical scheme are called *indistinguishable classical particles* if the random variables $\mathbf{X}$ are interchangeable, i.e. if

(7) $P(X_i = j_i, 1 \leq i \leq n) = P(X_{\pi(i)} = j_i, 1 \leq i \leq n)$

DEFINITION 3.3. For any configuration $\mathbf{j} \in \{1, \ldots, d\}^n$ we define the *occupation numbers* $\mathbf{k} \in \{0, 1, \ldots, n\}^d$ by

(8) $k_i(j) = \sum_{m=1}^{n} \delta_{i, j_m}$, $1 \leq i \leq d$,

and for the set of configuration random variables $\mathbf{X}$ we define the *occupation number random variables* $K_i: \Omega \rightarrow \{0, 1, \ldots, n\}$, $1 \leq i \leq$

d, by

(9) $K_i = \sum\limits_{m=1}^{n} 1_{[X_m = i]}$

where $1_{[X_m = i]}: \Omega \rightarrow \{0, 1\}$ denotes the indicator random variable of the event $[X_m = i]$.

Whenever the configuration random variables are given, the distribution of the occupation number random variables is uniquely determined. For interchangeable random variables the converse statement holds.

COMBINATORIAL LEMMA. *Assume that the random variables* X_i: $\Omega \rightarrow \{1, \ldots, d\}, 1 \leq i \leq n$, *are interchangeable then*

(10) $P(\mathbf{X} = \mathbf{j}) = \left(_{k_1(\mathbf{j})\,\ldots\,k_d(\mathbf{j})}^{\qquad n}\right)^{-1} P(\mathbf{K} = \mathbf{k}(\mathbf{j}))$.

Proof. Assume $\mathbf{j}, \mathbf{j}' \in \{1, \ldots, d\}^n$ such that $\mathbf{k}(\mathbf{j}) = \mathbf{k}(\mathbf{j}')$ holds. Then there exists a $\pi \in S_n$ such that $j_i = j'_{\pi(i)}, 1 \leq i \leq n$, and interchangeability implies that $P(\mathbf{X} = \mathbf{j}) = P(\mathbf{X} = \mathbf{j}')$. For any $\mathbf{j}$ there are $\left(_{k_1(\mathbf{j})\,\ldots\,k_d(\mathbf{j})}^{\qquad n}\right)$ different configurations with the same occupation number $\mathbf{k}(\mathbf{j})$. From this identity (10) follows.

We are prepared now for the definition of the familiar three statistics.

DEFINITION 3.4. Indistinguishable classical particles of the statistical scheme are called distributed according to
    (i) *Maxwell—Boltzmann statistics* if

(11) $P_{MB}(\mathbf{X} = \mathbf{j}) = d^{-n}$,

    (ii) *Bose—Einstein statistics* if

(12) $P_{BE}(\mathbf{X} = \mathbf{j}) = \left(_{k_1(\mathbf{j})\,\ldots\,k_d(\mathbf{j})}^{\qquad n}\right)^{-1} \binom{d+n-1}{n}^{-1}$,

    (iii) *Fermi—Dirac statistics* if $n \leq d$ and

(13) $P_{FD}(\mathbf{X} = \mathbf{j}) = (n!)^{-1} \binom{d}{n}^{-1}$ if $\mathbf{k} \in \{0, 1\}^d$

         $= 0$ if $\mathbf{k} \notin \{0, 1\}^d$.

Remarks. (1) According to our definition the particles of MB statistics are indistinguishable. As the trajectories do not enter into the description the particles cannot be distinguished by them. (2) In contrast to the traditional procedure we define the three statistics on the level of the d^n different configurations. Events that differ from one another by a permutation of the particles have the same probability and can, therefore, not be distinguished by their probabilities. In our setting events with more than one particle in a cell for FD statistics exist. (3) In quantum theory bosons, fermions and the particles of quantum MB statistics are three sets of states whereas these statistics, in the classical setting, are three states. (4) Due to the fact that in the traditional setting joint probabilities for the particles of BE and FD statistics are not defined, an analysis of statistical correlations is impossible. The bunching properties of BE and the antibunching properties of FD statistics can easily be deduced from our definitions.

Generalizing MB, BE, and FD statistics Brillouin [8] introduced a stochastic process which admits the construction of a large class of indistinguishable classical particles. For the formulation of Brillouin statistics we assume that a partition $D = (D_1, \ldots, D_f)$, $2 \leq f \leq d$, of the set $\{1, \ldots, d\}$ into f non-empty mutually disjoint subsets D_i, $1 \leq i \leq f$, is given such that $\cup D_i = D$ and $d_i = |D_i|$ satisfies $\Sigma d_i = d$. Moreover we introduce the random variables $Y_i: \Omega \rightarrow \{1, \ldots, f\}$, $1 \leq i \leq n$, where $[Y_i = j]$ denotes the event that particle i is in some cell of group j.

DEFINITION 3.5. The *Polya–Brillouin process* for the random variables $Y_i: \Omega \rightarrow \{1, \ldots, f\}$, $1 \leq i \leq n$, the partition D, specified by $(d_1, \ldots, d_f)$, and the parameter $c \in C$,

(14) $C = \{x \in \mathbb{R}; x \geq 0 \quad \text{or} \quad x = -1/m, m \in \mathbb{N}\}$,

is defined by the initial condition

(15) $P(Y_1 = j) = d_j/d, \qquad 1 \leq j \leq f$,

and the conditional probabilities, $2 \leq i \leq n - 1$.

(16) $P(Y_{i+1} = j | Y_1 = j_1, \ldots, Y_i = j_i) = \dfrac{d_j + cn_j(i)}{d + ci}$

where $n_j(i)$ is given by

(17) $n_j(i) = \displaystyle\sum_{m=1}^{i} \delta_{j, j_m}$.

From this definition we obtain by iteration the joint distribution of
$\mathbf{Y} = (Y_1, \ldots, Y_n), \mathbf{j} \in \{1, \ldots, f\}^n$,

$$
(18) \quad P(\mathbf{Y} = \mathbf{j}) = \frac{\prod_{i=1}^{f} d_i(d_i + c) \ldots (d_i + c(n_i - 1))}{d(d + c) \ldots (d + c(n - 1))}
$$

where $n_i = n_i(n)$. According to Equation (18) the random variables Y
are interchangeable and for $f = d$, $d_i = 1$, $1 \leq i \leq d$, we recover MB
($c = 0$), BE ($c = 1$) and FD ($c = -1$) statistics.

In analogy to Equations (8) and (9) we introduce occupation
numbers of groups $\mathbf{n} \in \{0, 1, \ldots, n\}^f$, $\Sigma n_i = n$, and occupation
number random variables of groups $N_i: \Omega \rightarrow \{0, 1, \ldots, n\}$, $1 \leq i \leq f$.
These random variables are distributed according to *Polya–Brillouin
distributions*

$$
(19) \quad P(\mathbf{N} = \mathbf{n}) = \binom{n}{n_1 \ldots n_f} \frac{\prod_{i=1}^{f} d_i(d_i + c) \ldots (d_i + c(n_i - 1))}{d(d + c) \ldots (d + c(n - 1))} .
$$

Due to the constraint $\Sigma N_i = n$ the random variables $\mathbf{N}$ are statistically
dependent. For any value of the parameter c, however, in the macro-
scopic limit n, $d \rightarrow \infty$, $n/d \rightarrow \bar{n}$ (where d_i, $1 \leq i \leq f - 1$, are fixed
and the f-th group acts as a reservoir) these random variables become
independent.

DEFINITION 3.6. For any sequence of occupation numbers $\mathbf{k} \in \{0,
1, \ldots, n\}^d$, $\Sigma k_i = n$, we define the *occupancy numbers* $\mathbf{z} \in \{0, 1, \ldots,
d\}^{n+1}$ by

$$
(20) \quad z_i = \sum_{m=1}^{d} \delta_{i, k_m}, \qquad 0 \leq i \leq n,
$$

and for the set of occupation number random variables $\mathbf{K}$ we define the
occupancy number random variables $Z_i: \Omega \rightarrow \{0, 1, \ldots, d\}$, $0 \leq i \leq
n$ by

$$
(21) \quad Z_i = \sum_{m=1}^{d} 1_{[K_m = i]}.
$$

According to Equation (19) the occupation number random variables K are interchangeable for any value of the parameter $c \leq C$. Therefore we obtain by means of the combinatorial lemma

$$(22) \quad P_{MB}(\mathbf{Z} = \mathbf{z}) = \frac{d!}{z_0! \ldots z_n!} \frac{n!}{0!^{z_0} \ldots n!^{z_n}} d^{-n},$$

$$(23) \quad P_{BE}(\mathbf{Z} = \mathbf{z}) = \frac{d!}{z_0! \ldots z_n!} \binom{d+n-1}{n}^{-1}.$$

We remark that some results concerning $P_{MB}(\mathbf{Z} = \mathbf{z})$ of von Mises [9] immediately extend to the distribution of the occupany number random variables for Brillouin statistics.

4. THE THEOREM OF DE FINETTI AND CORRELATION INEQUALITIES

From FD statistics it is well known that when the number of particles is equal to the number of cells, $n = d$, it is impossible to enlarge the number of particles without destroying the statistical properties of the system. This means that a sequence of interchangeable random variables is not necessarily extendible.

DEFINITION 4.1. Interchangeable random variables X_i, $1 \leq i \leq n$, are called N-extendible, $N > n$ and $N \in \mathbb{N}$ or $N = \infty$, if there exists a sequence of interchangeable random variables Y_i, $1 \leq i \leq N$ which satisfies for any $\mathbf{x} \in \mathbb{R}^n$

$$(24) \quad P(X_i < x_i, 1 \leq i \leq n) = P(Y_i < x_i, 1 \leq i \leq n).$$

CORRELATION LEMMA (cf. e.g. [7]). *Assume the random variables* X_i, $1 \leq i \leq n$, *are interchangeable, then the correlation coefficient satisfies the inequality*

$$(25) \quad Cor(X_1, X_2) \geq -\frac{1}{n-1}.$$

Remarks. (1) Equation (25) implies that an ∞-extendible sequence of interchangeable random variables necessarily has nonnegative correlations. Conversely, an interchangeable sequence with stricly negative correlations is not ∞-extendible. (2) A sequence of interchangeable random variables with nonnegative correlations is not necessarily ∞-extendible and if it is extendible at all the extension is not necessarily unique. (3) Obviously, the extendibility properties of a sequence of interchangeable random variables is intimately connected with the correlation structure. There are no general results how far a given interchangeable sequence is extendible. (4) The quantum analogue of the concept of N-extendibility is known as N-representability (cf. e.g. [10]). (5) For an ∞-extendible sequence of interchangeable random variables de Finetti's theorem holds.

THEOREM 4.1. (*de Finetti*) (*cf. e.g.* [7, 11]. *Assume the random variables* $X_i: \Omega \rightarrow \{0, 1\}$, $i \in \mathbb{N}$, *are interchangeable then there exists an uniquely determined probability measure* $v \in M_+^1([0, 1])$ *such that for any* $n \in \mathbb{N}$ *and for all* k, $0 \leq k \leq n$ *we have*

$$(26) \quad P\left(\sum_{i=1}^{n} X_i = k \right) = \int dv(p) \binom{n}{k} p^k (1 - p)^{n-k}.$$

Moreover, the mixing measure v *is determined by the Law of Large Numbers* (*LLN*)

$$(27) \quad \lim_{n \to \infty} E \exp\left(it \frac{1}{n} \sum_{i=1}^{n} X_i \right) = \int dv(p) \exp(itp).$$

Remarks. (1) For finitely extendible sequences of $\{0, 1\}$-valued random variables there always exists a representation as compound hypergenometric distribution. (2) The integral representations of de Finetti's theorem and its various generalizations are mixtures of distributions (states) of independent identically distributed particles. (3) According to Equations (26, 27) the mixing measure and the interchangeable sequence determine each other. For an ∞-extendible sequence of interchangeable random variables the local properties (26) are completely determined by the macroscopic properties (LLN).

For Brillouin statistics with negative values of the parameter c, $c = -1/m$, $m \in \mathbb{N}$, the associated interchangeable sequence is dm-extendible (but not $(dm + 1)$-extendible). This generalizes FD statistics insofar as a maximum of exactly m particles in a cell is admitted. For Brillouin statistics with nonnegative values of the parameter c the sequence is ∞-extendible by construction. In this case the multivariate version of de Finetti's theorem determines an integral representation for the Polya-Brillouin distributions (19) (cf. [12]).

In statistical mechanics one is usually interested in macroscopic properties. This situation is described by the Poisson limit of de Finetti's theorem.

THEOREM 4.2. [13]. *Let there be given random variables* $X_{n,i}:\Omega_n \to \{0, 1\}$, $i \in \mathbb{N}$, $n \in \mathbb{N}$, *defined for any n on a probability space* (Ω_n, F_n, P_n) *which are supposed to be interchangeable for any fixed n. Denote by* $\mu_n \in M^1_+(\mathbb{R}_+)$ *the image of the mixing measure* $\nu_n \in M^1_+([0, 1])$ *associated with the sequence* $X_{n,i}$ *(n fixed) under the mapping* $T_n: [0, 1] \to \mathbb{R}_+$ *defined by* $T_n(p) = np$. *Moreover, assume that the sequence* μ_n *converges weakly to a probability measure* $\mu \in M^1_+(\mathbb{R}_+)$ *then for any k* $\in \mathbb{Z}_+$

$$(28) \quad \lim_{n \to \infty} P_n \left(\sum_{i=1}^{n} X_{n,i} = k \right) = \int d\mu(x) \frac{1}{k!} x^k \exp(-x)$$

holds. Moreover, the mixing measure is uniquely determined by the following extension of the LLN

$$(29) \quad \lim_{n \to \infty} \lim_{m \to \infty} E_n \exp \left(it \frac{n}{m} \sum_{i=1}^{m} X_{n,i} \right) = \int d\mu(x) \exp(itx).$$

Remarks. (1) Assume a random variable $N: \Omega \to \mathbb{Z}_+$ is distributed according to a mixture of Poisson distributions, then this random variable has super-Poisson statistics, i.e.

$$(30) \quad \text{Var}(N) \geq E(N).$$

(2) The conditions for the convergence to a mixture of Poisson distributions in Theorem 4.2. are sufficient. Necessary conditions on the extendibility properties of an interchangeable array are given in [14]. (3) The integral representations by means of the Poisson limit of de Finetti's theorem for the macroscopic limit of Brillouin statistics are derived in [12].

For the generalization of de Finetti's theorem to the framework of quantum theory we need some notation. In the following $B(H)$ denotes the C^*-algebra of bounded linear operators on a separable complex Hilbert space H. By $\otimes^\mathbb{N} B(H)$ we denote the C^*-tensor product of countably many copies of $B(H)$. For details we refer the reader to [15].

THEOREM 4.3. [15]. *Assume E is a locally normal symmetric state on*

$\otimes^N B(H)$. *Then there exists an uniquely determined probability measure ν on the set of normal states on $B(H)$ such that for all $n \in \mathbb{N}$ the marginal state E_n defined on $\otimes^n B(H)$ admits the integral representation*

$$(31) \qquad E_n(\cdot) = \int d\nu(e)\,(\otimes^n e)\,(\cdot).$$

Moreover, if in addition the state is BE *symmetric the mixing measure ν is supported by the set of pure normal states on $B(H)$.*

Remarks. (1) FD symmetric states on the infinite tensor product do not exist. (2) According to Definition 2.3. the state $\otimes^n e$ on $\otimes^n B(H)$ is MB symmetric in general, and BE symmetric if e is a pure state on $B(H)$. (3) For an ∞-extendible interchangeable sequence the fundamental correlation inequality, $\mathrm{Cor}(X_1, X_2) \geq 0$, follows from the correlation lemma. The simplest quantum analog for a locally normal symmetric state E on $\otimes^N B(H)$ is, $b \in B(H)$,

$$(32) \qquad E_2(b \otimes b) - E_1^2(b) \geq 0.$$

(4) For $H = \mathbb{C}^2$ and the canonical representation of statistical operators by means of the Pauli matrices $\boldsymbol{\sigma}$

$$(33) \qquad w = \tfrac{1}{2}(1 + \mathbf{n} \cdot \boldsymbol{\sigma}),\, \mathbf{n} \in \mathbb{R}^3,\, |\mathbf{n}| \leq 1$$

we obtain under the conditions of Theorem 4.3., $\mathbf{a}, \mathbf{b} \in \mathbb{R}^3$,

$$(34) \qquad E_2(\mathbf{a} \cdot \boldsymbol{\sigma} \otimes \mathbf{b} \cdot \boldsymbol{\sigma}) = \int d\nu(\mathbf{n})\, \mathbf{a} \cdot \mathbf{n}\, \mathbf{n} \cdot \mathbf{b}.$$

Accordingly, in the sense of [16] we have, for marginal states of symmetric states on the infinite tensor product, a Kolmogorovian model such that classical inequalities can be used. From the inequality (cf. [16]), $\alpha, \beta, \gamma \in [-1, 1]$,

$$(35) \qquad |\alpha\beta - \beta\gamma| \leq 1 - \alpha\gamma$$

we immediately obtain the Bell type inequality, $|a| = |b| = |c| = 1$,

$$(36) \qquad |E_2(\mathbf{a} \cdot \boldsymbol{\sigma} \otimes \mathbf{b} \cdot \boldsymbol{\sigma}) - E_2(\mathbf{b} \cdot \boldsymbol{\sigma} \otimes \mathbf{c} \cdot \boldsymbol{\sigma})|$$
$$\leq 1 - E_2(\mathbf{a} \cdot \boldsymbol{\sigma} \otimes \mathbf{c} \cdot \boldsymbol{\sigma}).$$

A violation of this and equivalent inequalities implies that the symmetric state E_2 under consideration is not ∞-extendible.

As a last example of those integral representations of states which can be derived from de Finetti's theorem we consider the classical states of the quantum harmonic oscillator which are extensively used in quantum optics.

DEFINITION 4.2. Let $\Psi(z)$, where $z \in \mathbb{C}$, denote a coherent vector of the harmonic oscillator

(37) $\Psi(z) = \exp(za^+ - \bar{z}a)\,\Phi(0)$

where $\Phi(0)$ denotes the ground state. A state which can be identified with the statistical operator

(38) $W(\mu) = \int d\mu(z)\,P(z)$

where

(39) $P(z) = \langle \Psi(z), \cdot \rangle\,\Psi(z)$

and $\mu \in M^1_+(\mathbb{C})$ is called a *classical state*.

Remarks. (1) For the properties of classical states we refer to [17, 18]. For a derivation of classical states by means of a quantum version of the Poisson limit of de Finetti's theorem we refer to [19, 20]. The possibility of such a characterization of the classical states raised my interest in de Finetti's theorem and the underlying concepts. (2) For nonclassical states of the quantum harmonic oscillator certain correlation inequalities are violated. The properties of nonclassical photon states, e.g. antibunching and squeezing (cf. e.g. [21]), are the result of insufficient extendibility properties of the BE symmetric states on the array of two-level-systems (quanta) under consideration. This means that the nonclassical properties of nonclassical states are the result of specific correlations. (3) Whereas the P-representation of nonclassical states (cf. e.g. [22]) is based on concepts (distributions, signed measures) which have no physical meaning, the mixing measure μ of a classical state $W(\mu)$ has a precise physical meaning.

5. HISTORICAL AND CONCEPTUAL REMARKS

BE statistics was introduced, on the level of the occupancy numbers, by Boltzmann in 1877 [23] in connection with his probabilistic entropy concept

(40) $S_B \propto \max\{\ln P_{BE}(\mathbf{Z} = \mathbf{z})\}$, $\Sigma z_i = d$, $\Sigma i z_i = n$

which is equivalent to the entropy of the probability distribution for the normalized occupancy numbers $p_i = z_i/d, 0 \leqq i \leqq n$,

$$(41)   S \propto \max\left\{- \sum_{i=0}^{n} p_i \ln p_i\right\},   \Sigma p_i = 1,   \Sigma i p_i = n/d.$$

The physical idea underlying Boltzmann's choice of P_{BE} for $P(\mathbf{Z} = \mathbf{z})$ is the assumption of equal a priori probabilities for any energy distribution (distribution of the occupation numbers) of the system of n energy elements and d molecules which he consideres. From the most probable value of the occupancy numbers Boltzmann obtains, in the macroscopic limit, a geometric distribution for the most probable normalized occupancy numbers and, in a second step, in the continuum limit the Maxwell—Boltzmann distribution (exponential distribution) for the energy. In his memoir Boltzmann explicitly considers MB statistics (Equation (22)) as a counterexample for his theory.

Using the concept of occupancy numbers and following Boltzmann without performing the continuum limit, Natanson in 1911 [24] and Bose in 1924 [25] derived Planck's radiation law. In contrast to this method, Planck in 1900 [26] used the concept of occupation numbers and the entropy definition

$$(42)   S_P = - k \frac{1}{d} \ln P_{BE}(\mathbf{K} = \mathbf{k}).$$

Whereas Boltzmann's concept based on $P_{BE}(\mathbf{Z} = \mathbf{z})$ was confused with $P_{MB}(\mathbf{K} = \mathbf{k})$, Planck's concept (occupation numbers) was generalized to a coarse grained description (occupation numbers of groups) and eventually formulated as follows (cf. e.g. [8])

(43) $S_{Br} = \max\{k \ln W(\mathbf{n})\}$, $\Sigma n_i = n$, $\Sigma \varepsilon_i n_i = \varepsilon$

(ε denotes the energy) where, for Brillouin statistics the 'number of complexions' $W(\mathbf{n})$ is defined by

$$(44)   W(\mathbf{n}) = \prod_{i=1}^{f} \frac{d_i(d_i + c) \ldots (d_i + c(n_i - 1))}{n_i!} = \prod_{i=1}^{f} W(n_i).$$

Remarks. (1) As Equation (41) shows, Boltzmann's entropy definition is equivalent (up to a constant) to the entropy concept of information theory. (2) The multiplicative structure of $W(\mathbf{n})$ guarantees the additivity of the entropy, but is incompatible with the statistical dependence of the occupation number random variables (cf., however, the remark following Equation (19)). (3) The entropy of old quantum statistics, Equation (43), has the following properties. First, the entropy definition depends on the parameter of Brillouin statistics. Therefore this definition cannot be considered as a general concept. Second, $W(\mathbf{n})$ is in general, and in particular for MB statistics, no integer. Therefore it is impossible to interpret $W(\mathbf{n})$ as a 'number of realizations'. (4) Equation (44) shows that the division by $n!$, for $W(\mathbf{n})$, is a general structural property of Brillouin statistics and not confined to MB statistics.

To summarize, the application of a particluar entropy definition (which is restricted to Brillouin statistics) in combination with the (almost always impossible) interpretation of $W(\mathbf{n})$ as the 'number of possible distributions' eventually supported the idea to consider 'the number of possible distributions', and in particular

(45) $W_{BE}(\mathbf{n}) = \binom{d+n-1}{n}$, $W_{FD}(\mathbf{n}) = \binom{d}{n}$,

as an ontological assertion concerning the existing fundamental events. According to these assumptions a description of the system on the level of the d different configurations is impossible.

In this lecture we have studied the concept of indistinguishable classical particles and its quantum analog which agrees with the traditional concept of indistinguishability in quantum theory. The concept of classical and quantum indistinguishability is based on the indistinguishability (identity) of the probabilities for events which differ from one another by a permutation of the particles. From the puristic viewpoint it makes, therefore, no sense to call these particles indistinguishable. This is due to the fact that indistinguishability (in contrast to identity) is no property of these particles: it is a property of their state.

This entails that, according to the definition, the problem to decide whether particles are distinguishable or not is no problem of more precise empirical efforts and no problem of the quantum measurement process as argued in [27]. Identical particles are indistinguishable if it is certain that their state is symmetric.

According to the definition, indistinguishability is an extrinsic property. With respect to this statement, however, we advocate the following viewpoint. For a system of macroscopically many particles it is almost impossible to prepare the system in a state which is not symmetric as this requires the identification of some 10^{23} objects. According to this viewpoint a macroscopic system of identical particles can be idealized as a system of indistinguishable particles and indistinguishability can be considered as an intrinsic property. This is usually done in quantum theory where the state space $\otimes^n H$ is restricted to the BE/FD symmetric subspace.

In classical statistical mechanics the reduction of the state space, the space of all probability measures, to the subset of symmetric probability measures has drastic consequences. As in general the pure symmetric states do not generate the set of all symmetric states as their convex closure (cf. [28]), the classical ignorance interpretation of probability is no longer possible. Moreover, this property entails that indistinguishable classical particles are inherently indeterministic and it implies that indistinguishable classical particles have no trajectories. These properties, however, follow already from the definition because of the fact that, in general, identical classical particles can be identified in any deterministic situation. In other words, the concept of indistinguishable particles is a genuine probabilistic concept.

We conclude with the remark that by means of the identification of indistinguishability with permutation invariance of the state, the concept of indistinguishability is not exhausted. This concept presupposes that it is possible, in principle, to identify those events which cannot be distinguished by their probabilities.

Whereas in classical physics it is taken for granted that different events can always be distinguished, in quantum physics there exist different events which cannot, in principle, be distinguished. A well-known example for this concept of indistinguishability, which we would like to call indiscernibility, is the famous 2-slit experiment. Without destroying the experimental arrangement (and the result) it is impossible to distinguish the events 'the particle passes through hole 1' and 'the particles passes through hole 2'. The quantum rules for the description of this situation which is interrelated but not identical with the concept indistinugishable particles are explained by Feynman (cf. e.g. [29]).

BIBLIOGRAPHY

[1]   Jauch, J. M. 1968. *Foundations of Quantum Mechanics*. Reading: Addison-Wesley.

[2]   Girardeau, M. D. 1965. 'Permutation Symmetry of Many-Particle Wave Func-
      tions', *Phys. Rev.* **139** B500—B508.
[3]   Leinaas, J. M. and Myrheim, J. 1977. 'On the Theory of Identical Particles'.
      Nuovo Cimento **37B** 1—23.
[4]   Fleig, W. 1983. 'On the Symmetry Breaking Mechanism of the Strong-Coupling
      BCS-Model'. *Acta Physica Austriaca* **55** 135—153.
[5]   Ali, S. T. and Doebner, H. D. 1987. 'Quantization, Topology and Ordering. In
      The Physics of Phase Space, ed. by Y. S. Kim and W. W. Zachary, Lecture Notes
      in Physics **278**, Berlin: Springer, pp. 330—346.
[6]   Bach, A 1988. 'The Concept of Indistinguishable Particles in Classical and
      Quantum Physics'. *Found. Phys.* **18** 639—649.
[7]   Aldous, D. 1985. 'Exchangeability and Related Topics. In *Ecole d'Eté de
      Probabilités de Saint-Flour XIII*, P. L. Hennequin, Lecture Notes in Mathematics
      1117, Berlin: Springer, pp. 1—198.
[8]   Brillouin, L. 1931. *Die Quantenstatistik*. Berlin: Springer.
[9]   von Misés, R. 1964. 'Über Aufteilungs- und Besetzungswahrscheinlichkeiten'. In
      Selected papers of Richard von Mises, ed. by Ph. Frank *et al.*, Providence: AMS,
      pp. 313—334.
[10] Coleman, A. J. 1963. 'Structure of Fermion Density Operators'. *Rev. Mod. Phys.*
      35 668—689.
[11] Feller, W. 1966. *An Introduction to Probability Theory and its Applications*, Vol.
      2. New York: Wiley.
[12] Bach, A. 1988. 'Integral Representations for Indistinguishable Classical Particles:
      Brillouin Statistics. *Phys. Lett.d A* **129** 440—442.
[13] Benczur, A. 1968. 'On Sequences of Equivalent Events and the Compound
      Poisson Process'. *Stud. Sci. Math. Hung.* **2** 451—458.
[14] Bach, A. 1989. 'On the Statistics of Nonclassical Photon States'. *Phys. Lett. A*
      134 405—408.
[15] Hudson, R. L. and Moody, G. R. 1976. 'Locally Normal Symmetric States and
      an Analogue of de Finetti's Theorem'. *Z. Wahrscheinlichkeitstheorie verw.
      Gebiete* **33** 343—351.
[16] Accardi, L. *Foundations of Quantum Mechanics: A Quantum Probabilistic
      Approach*. Preprint, Rome.
[17] Bach, A. and Lüxmann-Ellinghaus, 1986. 'The Simplex Structure of the Classical
      States of the Quantum Harmonic Oscillator'. *Commun. Math. Phys.* **107** 553—
      560.
[18] Bach, A. 1988. 'Quanta and Coherent States', *Lett. Math. Phys.* **15** 75—79.
[19] Bach, A. 1987. 'Integral Representations by Means of Coherent States Derived
      from de Finetti's Theorem'. *Europhys. Lett.* **4** 383—387.
[20] Bach, A. and Srivastav, A. 1989. 'A Characterization of the Classical States of
      the Quantum Harmonic Oscillator by Means of de Finetti's Theorem'. *Commun.
      Math. Phys.* **123** 453—462.
[21] Walls, D. F. 1986. 'Quantum Statistics of Nonlinear Optics'. In *Frontiers of
      Nonequilibrium Statistical Physics*, ed. by G. T. Moore and M. O. Scully, New
      York: Plenum, pp. 309—328.
[22] Klauder, J. R. and Sudarshan, E. C. G. 1968. *Fundamentals of Quantum Optics*.
      New York: Benjamin.

[23]   Boltzmann, L. 1909. 'Über die Beziehung zwischen dem zweiten Hauptsatze der mechanischen Wärmetheorie und der Wahrscheinlichkeitsrechnung, respektive den Sätzen über das Wärmegleichgewicht'. In L. Boltzmann, *Wissenschaftliche Abhandlungen*, ed by F. Hasenöhrl Vol. 2. J. A. Barth, Leipzig, pp. 164—223.

[24]   Natanson, L. 1911. 'Über die statistische Theorie der Strahlung'. *Phys. Zs.* **12** 659—666.

[25]   Bose, S. N. 1924. 'Plancks Gesetz und Lichtquantenhypothese'. *Z. Phys.* **26** 178—181.

[26]   Planck, M. 1900. 'Zur Theorie des Gesetzes der Energieverteilung im Normalspektrum'. *Verh. d. Deutsch. Phys. Ges.* (2) **2** 237—245.

[27]   Maddox, J. 1987. 'Telling One Particle from Another'. *Nature* **329** 579.

[28]   Bach, A. 1985. 'On the Quantum Properties of Indistinguishable Classical Particles'. *Lett. Nuovo Cimento* **43** 383—387.

[29]   Feynman, R. P. 1961. *The theory of fundamental processes.* New York: Benjamin.

DOMENICO COSTANTINI AND UBALDO GARIBALDI

THE NON FREQUENCY APPROACH TO ELEMENTARY
PARTICLE STATISTICS

1. In 1927 L. Brillouin [1] deduced elementary particle statistics supposing distinguishability of all elementary particles and making use of a sort of geometrical probability related to capacities of cells and volumes of particles. In the present paper we show that Brillouin's approach can be restored without making any reference to the problem of distinguishability. In doing this, we refer to a probability concept which has nothing to do with relative frequency, but is explicitly related to single events.

To make clear the peculiarity of our approach, it is useful to recall that since the time of J. C. Maxwell and L. Boltzmann, elementary particle statistics has been conceived as distributions on possible populations. When particles are taken as distinguishable, the Maxwell–Boltzmann (MB) statistics is the appropriate distribution. When particles are considered as indistinguishable, the Bose-Einstein (BE) statistics is the appropriate distribution. When, moreover, two particles cannot occupy the same cell, the Fermi-Dirac (FD) statistics is the appropriate distribution.

Considering a population of two particles (members), X_1 and X_2, and three cells (predicates), 1, 2 and 3, the number of possible populations is:

in MB case, 9, i.e. $X_1 = 1$, $X_2 = 1$; $X_1 = 2$, $X_2 = 1$; ...; $X_1 = 3$, $X_2 = 3$;

in BE case, 6, i.e. $(2, 0, 0), (1, 1, 0), \ldots, (0, 0, 2)$;

in FD case, 3, i.e. $(1, 1, 0), (1, 0, 1), (0, 1, 1)$.

The MB statistics allots the value $1/9$ to each of the nine possible population; the BE statistics the value $1/6$ to each of the six possible populations; the FD statistics the value $1/3$ to each of the three possible populations. This is the standard way of facing the three familiar statistics.

A second approach makes use of the product rule of probability. Let us consider a possible population in the MB case, for example, $X_1 = 1$, $X_2 = 2$. By the product rule one has

$$\Pr\{X_1 = 1, X_2 = 2\} = \Pr\{X_1 = 1\}\Pr\{X_2 = 2 \mid X_1 = 1\}.$$

R. Cooke and D. Costantini (eds), Statistics in Science. The Foundations of Statistical Methods in Biology, Physics and Economics, 167–181.
© 1990 *Kluwer Academic Publishers.*

That is, the probability of populations can be derived as the product of probabilities describing accommodations of particles in cells. As a consequence, knowing the probability with which a particle occupies a cell, given each possible state of the system, one can determine the probability of each possible population. Combinatorial techniques easily give the probability of each possible population in BE and FD cases too.

Formally, the two approaches are completely equivalent. However, philosophies behind them are quite different. The traditional approach rests on the frequency interpretation of probability, whilst the second is more congenial with a non frequency point of view. Moreover, from first view point it seems impossible to use the second approach when particles are indistinguishable. As a matter of fact this is not the case as will be seen later. However, before considering this approach in some details, it is worth giving a brief history of it.

2. In 1924 [2] W. E. Johnson proposed a formula for determining the probability that, in a population of n members, an unobserved case has the predicate j, $1 \leqslant j \leqslant d$, given any supposed vector (sample distribution) $\mathbf{N}_s = \mathbf{s} = (s_1, \ldots, s_d)$, $\Sigma_j s_j = s$, of s examined cases. In modern notation the formula is as follows

$$(1) \qquad \Pr\{X_i = j \mid \mathbf{N}_s = \mathbf{s}\} = \frac{1 + s_j}{s + d}, \qquad i > s.$$

In order to derive (1) Johnson assumed a permutation postulate (symmetry or exchangeability) and a combination postulate: all structurally similar populations are equiprobable.

In 1925 (see the note of R. B. Braithwait in [3]) Johnson abandoned the combination postulate for a sufficiency postulate: the probability that the next case will bear j depends only on s_j and s. From this assumption, Johnson deduced the following equality

$$(2) \qquad \Pr\{X_i = j \mid \mathbf{N}_s = \mathbf{s}\} = \frac{1 + w s_j}{d + w s}, \qquad i > s$$

where w is a non negative constant. It is easy to deduce from (2), via the product rule, the probability of all possible populations in the MB, BE, and FD cases.

In 1927, assuming particles, whether classical or not, to be distinguishable and considering a cell C_j with g_j levels, $\Sigma_j g_j = G$, Brillouin used the product rule to determine probabilities of configurations

$$C = \{X_1 = j_1, \ldots, X_n = j_n\}$$

and of the corresponding occupation vector

$$\mathbf{n} = (n_1, \ldots, n_d); \; n = \sum_{j=1}^{d} n_j.$$

In order to reach his basic formula, Brillouin supposed that each particle has a volume equal to a; each void level has a capacity equal to 1; if a level contains s_j particles, its capacity becomes $1 - as_j$.

From this assumption one can derive the probability of putting the first particle in one level of C_j, i.e.

$$\Pr\{X_1 = j\} = g_j/G.$$

The probability of putting the $(s + 1)$-th particle in C_j is

$$(3) \qquad \Pr\{X_{s+1} = j \mid \mathbf{N}_s = \mathbf{s}\} = \frac{g_j - as_j}{G - as}.$$

In terms of Euler's Γ-function, the probability of a configuration is obtained from (3)

$$(4) \qquad \Pr\{C\} = \prod_j a^{n_j} \frac{\Gamma(g_j/a + 1)}{\Gamma(g_j/a - n_j + 1)} \frac{\Gamma(G/a - n + 1)}{a^n \Gamma(G/a + 1)}.$$

From (4) it follows that all configurations belonging to the same occupation vector are equiprobable. Hence

$$(5) \qquad \Pr\{\mathbf{n}\} = n! \prod_j \frac{a^{n_j}}{n!} \frac{\Gamma(g_j/a + 1)}{\Gamma(g_j/a - n_j + 1)} \frac{\Gamma(G/a - n + 1)}{a^n \Gamma(G/a + 1)}.$$

Finally, Brillouin obtained MB, BE, and FD statistics putting in (5) $a = 0$, $a = -1$ and $a = +1$ respectively.

Brillouin's suggestion was not accepted, and his unified theory was forgotten. One possible, but not the sole reason, for this is the philosophy implicitly assumed by his basic formula (3). This formula gives the probability with which a particles goes into a cell. In other words, Brillouin, aiming to derive particle statistics, was using a concept of probability which has nothing to do with relative frequency. On the contrary, it is a conditional concept — the probability is relative to the state of the system — which explicitly refers to a single event, namely $X_{s+1} = j$, i.e. the event that the $(s + 1)$-th particle goes into the cell C_j.

This was in patent contrast with the heavy frequentistics atmosphere of the twenties.

In 1952 R. Carnap [4] having obtained, independently from Johnson and Brillouin, the formula

$$(6) \qquad \Pr\{X_i = j \mid \mathbf{N}_s = \mathbf{s}\} = \frac{\lambda p_j + s_j}{\lambda + s}, \qquad i > s$$

where $0 < \lambda \leqslant \infty$ and $p_j = \Pr\{X_i = j\}$, used it systematically for determining probabilities of state descriptions (configurations) and structure descriptions (occupation vectors).

In 1971 Carnap [5] gave a satisfactory theory of predictive probability functions, that is of function like (2), (3), and (6). It is worth stressing that, although Carnap worked only on populations with distinguishable members, his theory is based on a set of axioms referring to occupation vectors only.

In 1983 D. Costantini, M. C. Galavotti, and R. Rosa [6], considered all particles as distinguishable and applied Carnap's theory to statistics. Their approach is based on the notion of relevance quotient [7] at i, defined as the ratio between the transition probability to j at the $(i + 1)$-th step, provided that the $(i + 1)$-th particle is in g, $g \neq j$, and the transition probability to j at the i-th step.

The relevance quotient at $\mathbf{0}$, η (for the definition of this concept see (10)) enables the introduction of various types of correlation among particles. It is worth noting that for the parameter in (5), $\lambda = \eta/(1 - \eta)$ holds.

The three familiar statistics are obtained calculating the probability of occupation vectors, and then putting:

$\eta = 1$, for the MB statistics;

$\eta = d/(d + 1)$, for the BE statistics;

$\eta = d/(d - 1)$ for the FD statistics.

In this approach elementary particle statistics are obtained as the result of a stochastic process whose transition probabilities are given by (6).

In 1988 A. Bach [8] discovered Brillouin's work on particle statistics. Bach rightly pointed out that (5) is a family of multivariate Polya distributions. He called these distributions Polya—Brillouin distributions and named the process used to reach them, the Polya—Brillouin process. This is a process for the random variables X_s, $1 \leqslant s \leqslant n$, which is defined by the initial condition $\Pr\{X_s = j\} = g_j/G$ and the joint conditional probabilities

(7) $\quad \Pr\{X_{s+1} = j \mid X_1 = j_1, \ldots, X_s = j_s\} = \dfrac{g_j + cs_j}{G + cs}$,

$c \in C = \{x \in R; x \geqslant 0 \text{ or } x = -1/m, m \in N\}$ is the parameter of the process.

To the best of our knowledge, the last event of this story is the strictly probabilistic reformulation of Brillouin's unified theory [9]. This reformulation is carried out in terms of predictive functions of Johnson and Carnap. In the next section, we recall the main features of this reformulation.

3. If we consider n particles and d cell, the integers n_j, $1 \leqslant j \leqslant d$, $\Sigma_j n_j = n$, are the occupation numbers of the cells and $\mathbf{n} = (n_1, \ldots, n_d)$ is the occupation vector of the system. $^n N^d = \{\mathbf{n} \in N^d; \Sigma_j n_j = n\}$ is the set of all occupation vectors. This set has $\binom{n+d-1}{n}$ members. If, for the particles we are considering, the exclusion principle holds, the occupation numbers can only be 0 or 1. In this case, if $A = \{0, 1\}$, $^n A^d = \{\mathbf{n} \in A^d; \Sigma j\, n_j = n\}$ is the set of all occupation vectors. This set has $\binom{d}{n}$ members.

Let us consider a system with $s - 1$ particles, $s < n$. The s-th experiment is that which adds one particle to the system. Thus the s-th experiment is a decomposition of the sure event U, i.e.

$$U = E_s^1 \cup \ldots \cup E_s^d$$

where E_s^j is the event that takes place when the s-th experiment increase the occupation number of the j-th cell. Hence if E_s^j takes place, the occupation vector goes from $\mathbf{s}^{-j} = (s_1, \ldots, s_j - 1, \ldots, s_d)$ to $\mathbf{s} = (s_1, \ldots, s_j, \ldots, s_d)$, i.e. the occupation number of the j-th cell goes from $s_j - 1$ to s_j.

We denote with Y_i^j the indicator random variable of E_i^j. $(Y_i^1, \ldots, Y_i^d)$ is the random vector of the i-th experiment.

$$\mathbf{N}_s = (N_s^1, \ldots, N_s^d) = \sum_i (Y_i^1, \ldots, Y_i^d)$$

is the sum of the random vector up to the s experiment.

If in the $(s + 1)$-th experiment the event E_{s+1}^j occurs, then the occupation number of the j-th cell goes from s_j to $s_j + 1$ and thus the occupation vector goes from $\mathbf{s}$ to $\mathbf{s}^j = (s_1, \ldots, s_j + 1, \ldots, s_d)$. We name this the transition from $\mathbf{s}$ to $\mathbf{s}^j$. Let

$$(8) \qquad \Pr\{j|\mathbf{s}\} = \Pr\{Y^j_{s+1} = 1 | \mathbf{N}_s = \mathbf{s}\}$$

be the probability of this transition. (8) is a relative probability; more precisely, it is the probability that E^j_{s+1} occurs when the occupation vector is $\mathbf{s}$. This probability is symmetric when, for $j \neq g$

$$\Pr\{j|\mathbf{s}\}\Pr\{g|\mathbf{s}^j\} = \Pr\{g|\mathbf{s}\}\Pr\{j|\mathbf{s}^g\}$$

holds.

The notion of symmetric probability function can be formalized in the following way. A function $\Pr\{\cdot|\cdot\}$ with domain $\{1, \ldots, d\} \times {}^s N^d$ and codomain R is a symmetric probability function if, and only if, it satisfies the following conditions.

C1.   for each j, $\Pr\{j|\mathbf{s}\} \geq 0$.

C2. $\Sigma_j \Pr\{j|\mathbf{s}\} = 1$.

C3. $\Pr\{j|\mathbf{s}\}\Pr\{g|\mathbf{s}^j\} = \Pr\{g|\mathbf{s}\}\Pr\{j|\mathbf{s}^g\}, \qquad j \neq g$.

Let us now consider an allowed sequence of occupation vectors

$$(9) \qquad \{\mathbf{s}; 0 \leq s \leq n\}.$$

The first term of (9) is $\mathbf{0} = (0, \ldots, 0)$, the second is $\mathbf{0}^j \in {}^1 N^d$; in general, the $(s + 1)$-th term of the sequence is $\mathbf{s}^j \in {}^{s+1} N^d$; and the last term of (9) is $\mathbf{n}$. We name a sequence like (9) a pattern from $\mathbf{0}$ to $\mathbf{n}$ or, more simply, a pattern to $\mathbf{n}$. Thus a pattern is a conjunction of events like

$$U \cap E^{j_1}_1 \cap \ldots \cap E^{j_n}_n, \qquad 1 \leq j_i \leq d$$

that can be represented as

$$(\ldots(\mathbf{0}^{j_1})\ldots)^{j_n}, \qquad 1 \leq j_i \leq d.$$

There are $n!/\Pi_j n_j!$ patterns to $\mathbf{n}$. The main pattern to $\mathbf{n}$ is the pattern which puts n_1 particles first in the 1-th cell, then n_2 particles in the 2-th cell, $\ldots$, and finally n_d particles in the d-th cell. For fermions there are $n!$ pattern to $\mathbf{n}$. With the suitable limitations, the main pattern has the meaning we have just seen.

If, as in Section 1, we consider 3 cells and 2 particles, the occupation vectors are $(2, 0, 0)$, $(0, 2, 0)$, $(0, 0, 2)$, $(1, 1, 0)$, $(1, 0, 1)$, $(0, 1, 1)$. There is only one pattern to each of the first three vectors and two patterns to each of the other three, for example the patterns to $(1, 1, 0)$ are $\{(0, 0, 0), (1, 0, 0), (1, 1, 0)\}$, which is the main pattern, and $(0, 0, 0), (0, 1, 0), (1, 1, 0)\}$. If the two particles are fermion, the occupation vectors are only $(1, 1, 0)$, $(1, 0, 1)$ and $(0, 1, 1)$, and there are two patterns to each of them.

A pattern is a sequence of transitions and when we are able to determine the value of (8), we can also determine the probability of a pattern. C1, C2, and C3 are not enough to determine the value of (8), but from them follows

THEOREM 1. *If* Pr *is a symmetric probability function, then* Pr *allots the same probability to all patterns to the same* **n**.

The proof of this theorem, and of that of all other theorems in this paper are given in [9].

4. In order to determine the value of (8) we must introduce new conditions. The first of these is prior equiprobability.

C4.   for each j, $\Pr\{j \mid \mathbf{0}\} = d^{-1}$.

The other conditions, on ground of expediency, will be expressed using the notion of relevance quotient at **s** that we now define

$$Q_j^g(\mathbf{s}) = \Pr\{j \mid \mathbf{s}^g\}/\Pr\{j \mid \mathbf{s}\}, \qquad j \neq g.$$

Obviously, in order that this definition makes sense, $\Pr\{j \mid \mathbf{s}\}$ must be different from zero. From now on we will suppose that this is always the case. A particular case of special importance, that we already considered in Section 2, is the relevance quotient at **0**

(10) $Q_j^g(\mathbf{0}) = \Pr\{j \mid \mathbf{0}^g\}/\Pr\{j \mid \mathbf{0}\} = \eta.$

The condition we now consider determines the invariance of the relevance quotient. A Pr which is symmetric and satisfies the condition

C5.   if $g \neq j, q \neq f, \mathbf{s}', \mathbf{s}'' \in {}^sN^d(\mathbf{s}', \mathbf{s}'' \in {}^sA^d),$
        then $Q_j^g(\mathbf{s}') = Q_q^f(\mathbf{s}''),$

is called invariant.

We can now formulate some important results.

THEOREM 2. *If the particles we are considering satisfy the exclusion principle and* Pr *is an invariant, symmetric and prior equiprobable probability function, then*

(11) $\Pr\{j|\mathbf{s}\} = (d - s)^{-1}.$

Hence for fermions and for an invariant, symmetric and prior equiprobable function, the relevance quotient at $\mathbf{s}$ is equal to $(d - s)/(d - s - 1)$ and the relevance quotient at $\mathbf{0}$ is equal to $d/(d - 1)$.

THEOREM 3. *If the particles we are considering do not satisfy the exclusion principle and* Pr *is an invariant, symmetric and prior equiprobable probability function, then*

(12) $\Pr\{j|\mathbf{s}\} = \dfrac{\lambda/d + s_j}{\lambda + s}$

where $\lambda = \eta/(1 - \eta).$

As (12) shows, in the case of particles not satisfying the exclusion principle, the conditions we have considered are not enough to calculate the probability of a transition from $\mathbf{s}$ to $\mathbf{s}^j$. To this end we must choose a value of λ and thus of η.

We introduce two new conditions

C6'. $\eta = d/(d + 1),$

C6". $\eta = 1.$

We have

THEOREM 4. *If the particles we are considering do not satisfy the exclusion principle, and* Pr *is an invariant, symmetric and prior equiprobable probability function for which* C6' *holds, then*

(13) $\Pr\{j|\mathbf{s}\} = \dfrac{1 + s_j}{s + d}.$

THEOREM 5. *If the particles we are considering do not satisfy the exclusion principle, and* Pr *is an invariant, symmetric and prior equiprobable probability function for which* C6" *holds, then*

(14) $\Pr\{j|\mathbf{s}\} = 1/d.$

It is worth noting that for these probability functions the relevance quotients are respectively $(s + d)/(s + d + 1)$ and 1.

Using these theorems we can achieve the three familiar statistics. To this end, we firstly calculate the probability of the main pattern to $\mathbf{n}$, denoted by ^mn, assuming respectively the values 1, $d/(d + 1)$ and $d/(d - 1)$ for η.

(15)   If $\eta = 1$,   then $\Pr\{^mn\} = \prod_j d^{n_j} = d^{-n}$.

(16)   If $\eta = d/(d + 1)$,   then $\Pr\{^mn\} = \prod_j n_j! \left/ \prod_{i=0}^{n-1} (i + d) \right.$.

(17)   If $\eta = d/(d - 1)$,   then $\Pr\{^mn\} = \prod_{i=0}^{n-1} (d - i)^{-1}$.

In order to produce the three statistics it is enough to calculate the probability of an ocupation vector. Using (15) and (16) for $\mathbf{n} \in {}^nN^d$ we obtain the MB and the BE statistics respectively.

$$\Pr\{\mathbf{n}\} = \frac{n!}{\prod_j n_j!} \prod_i d^{-n_j} = \frac{n!}{\prod_j n_j!} d^{-n},$$

$$\Pr\{\mathbf{n}\} = \frac{n!}{\prod_j n_j!} \frac{\prod_j n_j!}{\prod_{i=0}^{n-1} (i + d)} = \binom{n+d-1}{n}^{-1}.$$

Using (17) for a vector $\mathbf{n} \in {}^nA^d$ we obtain the FD statistics

$$\Pr\{\mathbf{n}\} = \frac{n!}{\prod_{i=0}^{n-1} (d - 1)} = \binom{d}{n}^{-1}.$$

If we consider k cells endoved with levels $l_1, \ldots, l_k$, $\Sigma_i\, l_i = d$, and we work out the probability of occupation vectors relative to each cell, we obtain

$$\mathrm{Pr}_{\mathrm{MB}}\{\mathbf{n}\} = d^{-n} n! \prod_i (l_i^{n_i}/n_i!),$$

$$\mathrm{Pr}_{\mathrm{BE}}\{\mathbf{n}\} = \binom{d+n-1}{n}^{-1} \prod_i \binom{l_i+n_i-1}{n_i},$$

$$\mathrm{Pr}_{\mathrm{FD}}\{\mathbf{n}\} = \binom{d}{n}^{-1} \prod_i \binom{l_i}{n_i},$$

that is the MB, the BE and the FD distributions.

We have produced the three familiar statistics without using names for particles. The probability function we have used during our analysis only refers to cells, levels and occupation numbers. This means that the fact that the considered particles are or are not distinguishable is completely irrelevant to the deduction of the distribution ruling the statistical behaviour of the system. But this does not mean that particles in the system are indistinguishable. We shall return to this problem in the next section.

Our analysis shows that the value of η characterizes the probability functions (11), (13), and (14). It is easy to realize that η is a way of expressing the correlation among particles. More exactly: if $\eta > 1$, the correlation is negative; if $\eta = 1$, there is no correlation; if $\eta < 1$, the correlation is positive. Thus what really distinguishes the three types of particles is the correlation existing among them.

Finally, our characterization clearly shows that negative correlation among fermions ensues from the exclusion principle. In fact, C1, C2, C3, C4, and C5 are compatible with every value of η. As we have show, these conditions lead to (12), from which (11) can also be obtained: is sufficient to put $\lambda = -d$ in (12). It is the exclusion principle that leads to (11) and hence to negative correlation.

However, on the contrary, positive correlation for bosons and independence for classical particles must be explicitly put in. This is the specific task of C6′ and C6″. In other words, whilst negative correlation among fermions ensue from Pauli's exclusion principle, i.e. from the impossibility for two particle to accommodate in the same level, positive correlation do not follow from the mere possibility of accommodating more particles in the same level.

5. As we have seen in the preceding section, elementary particle statistics can be obtained from purely probabilistic hypotheses. There is no need to use non-probabilistic argument apart from Pauli's exclusion principle. Of course, it is possible to state this principle in probabilistic terms. For example: when a level contains one particle, then the probability that another particle goes in it is zero. But it is easy to realize that the impossibility stated by the principle is not probabilistic in character. With this exception, one can conclude that conditions which have been considered do not talk about particles, but about probabilities. More precisely, they talk about probabilities that particles occupy some energy levels.

This bears some consequences as to the problem of indistinguishability of elementary particles. As we already noted, no considerations regarding distinguishability or indistinguishability have been made in the preceding section. From this it follows that such considerations are not essential ingredients of the derivation of particle statistics. On the contrary, the notion we have extensively used is that of symmetry. This suggests that this notion should be analysed in some detail in order to investigate the relationship between indistinguishability and symmetry.

Symmetry tells us that probability does not change as the individuals taken into account change, provided that their number remains the same. A simple case in point is throwing a coin, say five time. If we consider two possible results of this group of throwings, for example $H_1 H_2 H_3 H_4 T_5$ and $H_1 T_2 H_3 H_4 H_5$, where H and T stand for Heads and Tails, a probability which is symmetric assigns them the same probability. The same holds for all five results in which T occurs just once. What this amounts to is a characteristic of the probability function. In no way does it imply that the throwings which have been made are not distinguishable. As a matter of fact, the five throwings could result from coins of different countries, and in this case one would at any moment be able to recognize which coins have given Heads or Tails. What symmetry implies is that, for the purpose of describing the set of throwings, it is useless to know the order in which the results have occurred.

A further example is given by an insurance company which intends to calculate the premium to be paid for taking out a life insurance policy for next year in the town of Genoa. Obviously the company will have to distinguish among people according to their age, sex, the fact that they are smokers or non-smokers, and so on, since the probability of death varies for such groups of people. However, once one group has been selected, for example, men born in 1940 who smoke, its components are considered as indistinguishable for the purpose of determining their probability of death. In such a group one determines the number of members at the beginning and at the end of the 49-th year of age: the relative frequency of death gives the probability sought. One can see that for the sake of determining this probability it is irrelevant whether Mr. Rossi has survived while Mr. Bianchi has died, or the converse. What matters is that, given that Mr. Bianchi has died, if we could bring him back from the dead, we would have to force Mr. Rossi to die prematurely. Such an exchange will possibly be relevant for Mr. Bianchi, but will be of no importance at all for the purpose of determining the premium to be paid to the insurance company.

This follows from the fact that the probability function used to determine the probability of death relative to a given group of men is symmetric, that is it does not change with respect to interchange in the individuals belonging to the same group. We could express this as well by saying that for the purpose of determining their probability of death, men from Genoa are indistinguishable. But would this mean that they cannot be distinguished? The obvious answer is that they are in fact perfectly distinguishable, even though they are considered indistinguishable by the probability used by the insurance company.

We are convinced that what we have said is connected also with the problem of the distinguishability of elementary particles. Obviously, coins and men are not particles, and while we can distinguish among men and coins of different countries we cannot distinguish among particles, since we cannot mark them. However, if all we have in order to describe the characteristic of particles is the macroscopic behaviour of systems of particles as derived from probabilistic assumptions, like symmetry, then nothing can be said about indistinguishability of particles. In other words, if all we know about a system is that it behaves according to Boltzmann's distribution, no conclusion can be drawn to the effect that particles are distinguishable or indistinguishable. It is trivially known that a gas of 'indistinguishable' quantum particles at high temperatures and low densities is ruled by Boltzmann's distribution. However, in the last section we have seen that indistinguishability can be dispensed with for the purpose of deriving MB statistics.

To sum up, symmetry has no bearing at all on indistinguishability.

On the other hand, as we have seen to derive the three familiar statistics there is no need to make use of the notion of distinguishability as well as that of indistinguishability. Statistics can be obtained from purely probabilistic assumptions, without appelling to such properties of particles.

6. In Section 2 we suggested one reason for Brillouin's attempt was disregarded. Now, after having faced the relationship between symmetry and indistinguishability, we point out another one. Basically, Brillouin tried to substitute the fuzzy picture of indistinguishable independent quantum particles with the more classical notion of correlated distinguishable particles. The reformulation, which suppose the permutational invariance of probability of configurations being the modern formulation of indistinguishability, take Brillouin's approach as describing distinguishable [6] or indistinguishable [8] correlated objects. However, it is noteworthy that the event $X_s = j_s$, which appears in (3), (6), and (7), is by no means observable if we are dealing with indistinguishable particles. As a consequence all approaches which rest on this sort of events would probably encounter the same difficulty as Brillouin's original paper.

Moreover, it must be observed that the configuration stochastic process

$$\Pr\{C\} = \Pr\{X_1 = j_1\}\Pr\{X_2 = j_2 \mid X_1 = j_1\}\ldots$$
$$\Pr\{X_n = j_n \mid X_1 = j_1, \ldots, X_{n-1} = j_{n-1}\}$$

ruled by the transition probability (7), admits only an epistemic interpretation. In fact, in this process one first looks for the level occupied by particle 1, then one studies particle 2 conditioned on the accommodation of particle 1, and so on, all particles being already accommodated.

If particles are indistinguishable, then this process is trivially symmetric, and it is not surprising that this hold independently of any physical constraint. Configurations could have a direct physical meaning only if particles were admitted into the system one at a time, and their naming would occur at the entrance of each particle in such a way that the i-th incoming particle is described by X_i. However, this naming procedure is misleading, as it mistakes the (supposed existing) name of the particle for the entrance order, that is an unquestionable physical notion.

On the contrary, events of type $Y_s^j = 1$ considered in (8) have a direct nontrivial physical meaning. They describe the increments of the occupation vector of a system that is growing adiabatically one particle at time, so that it can be always considered in equilibrium. If correlation eixsts among particles, then what may discriminate different identical particles is the order of entrance in the system.

Unfortunately, Brollouin's assumption of d equiprobable levels implies that the Y_s^j are symmetric, i.e. the entrance order is irrelevant and all patterns to the same $\mathbf{n}$ are equiprobable. Due to the fact that the number of distinct patterns to $\mathbf{n}$ are equal to number of distinct configurations belonging to the same $\mathbf{n}$, these two quite different notions are treated in the same way.

This confusing equality is removed when energy is considered. If we do it via the standard conditioned maximum method, the symmetry breaking is not so perspicous. A deeper insight is obtained applying to patterns the method of the canonical distribution

$$\Pr\{\mathbf{n}\} = Z_n^{-1} \gamma(\mathbf{n}) \exp\left(-\beta \sum_j n_j \varepsilon_j\right)$$

where Z_n is the n-particles partition function, $\gamma(\mathbf{n})$ the multiplicity of the state, ε_j, $j = 1, 2, \ldots$, the energy of the j-th level and β is the inverse temperature. It is easy to show [10] that the canonical distribution preserves symmetry of Y_s^j's only in the MB case, otherwise correlation and temperature destroy symmetry. In other words, a fermion at low temperature preferably occupies the lowest free level, that is a strong function of the number of already present particles. Hence, if two patterns to the same $\mathbf{n}$ differ in the order of two transitions, the pattern which first attains the lower level is more probable than the other one. For bosons the opposite occurs. On the contrary, the configuration process conserves all properties, but it is useless for deriving $\Pr\{\mathbf{n}\}$.

Coming back to Brillouin's paper, the approach clearly deals with effective volume of levels rather than distinguishability of particles. We have show that Brillouin's approach can be vindicated in a formal calculation. This does not imply that his effort, if it had been widely accepted, would have solved the question for all. A pure statistical approach simply aims at the best phenomenological description of the system, on the basis of clearly stated probabilistic principle. Quantum correlations can be accepted as primitive terms, or described by non-local action at distance corresponding to quantum potential for many body forces. But they cannot be ignored, as in the traditional approach of the so-called statistical mechanics of indistinguishable, independent non interacting particles.

Universitá di Genova, Genova, Italy

BIBLIOGRAPHY

[1]   Brillouin, L. 1927. 'Comparaison des differents statistiques appliquee aux problemes de Quanta'. *Annales de Phisique* **VII** 315—331.
[2]   Johnson, W. E. 1924. *Logic. Part III.* Cambridge: Cambridge University Press.
[3]   Johnson, W. E. 1932. 'Probability: The Deductive and Inductive Problems'. *Mind* **164** 409—423.
[4]   Carnap, R. 1952. *The Continuum of Inductive Methods.* Chicago: The University of Chicago Press.
[5]   Carnap, R. 1971. 'A Basic System of Inductive Logic'. *Studies in Inductive Logic and Probability*, ed. by Carnap and Jeffrey. Berkeley: University of California Press.
[6]   Costantini, D., Galavotti, M. C., and Rosa, R. 1983. 'A Set of 'Ground-Hypotheses' for Elementary-Particle Statistics'. *Il Nuovo Cimento* **74** B(2) 151—158.
[7]   Costantini, D. 1979. 'The Relevance Quotient'. *Erkenntnis* **14** 149—157.
[8]   Bach, A. 1988. 'Integral Representation for Indistinguishable Classical Particle: Brillouin Statistics'. *Physics Letters A* **129** 440—442.
[9]   Costantini, D. and Garibaldi, U. 1989. 'Classical and Quantum Statistics as Finite Random Processes'. *Foundation of Physics* **19** 743—754.
[10]   Costantini, D. and Garibaldi, U. (work in progress).

NAME INDEX

BOSTON STUDIES IN THE PHILOSOPHY OF SCIENCE

Editors:

ROBERT S. COHEN and MARX W. WARTOFSKY

(Boston University)

1. Marx W. Wartofsky (ed.), *Proceedings of the Boston Colloquium for the Philosophy of Science 1961–1962*. 1963.
2. Robert S. Cohen and Marx W. Wartofsky (eds.), *In Honor of Philipp Frank*. 1965.
3. Robert S. Cohen and Marx W. Wartofsky (eds.), *Proceedings of the Boston Colloquium for the Philosophy of Science 1964–1966. In Memory of Norwood Russell Hanson*. 1967.
4. Robert S. Cohen and Marx W. Wartofsky (eds.), *Proceedings of the Boston Colloquium for the Philosophy of Science 1966–1968*. 1969.
5. Robert S. Cohen and Marx W. Wartofsky (eds.), *Proceedings of the Boston Colloquium for the Philosophy of Science 1966–1968*. 1969.
6. Robert S. Cohen and Raymond J. Seeger (eds.), *Ernst Mach: Physicist and Philosopher*. 1970.
7. Milic Čapek, *Bergson and Modern Physics*. 1971.
8. Roger C. Buck and Robert S. Cohen (eds.), *PSA 1970. In Memory of Rudolf Carnap*. 1971.
9. A. A. Zinov'ev, *Foundations of the Logical Theory of Scientific Knowledge (Complex Logic)*. (Revised and enlarged English edition with an appendix by G. A. Smirnov, E. A. Sidorenka, A. M. Fedina, and L. A. Bobrova). 1973.
10. Ladislav Tondl, *Scientific Procedures*. 1973.
11. R. J. Seeger and Robert S. Cohen (eds.), *Philosophical Foundations of Science*. 1974.
12. Adolf Grünbaum, *Philosophical Problems of Space and Time*. (Second, enlarged edition). 1973.
13. Robert S. Cohen and Marx W. Wartofsky (eds.), *Logical and Epistemological Studies in Contemporary Physics*. 1973.
14. Robert S. Cohen and Marx W. Wartofsky (eds.), *Methodological and Historical Essays in the Natural and Social Sciences. Proceedings of the Boston Colloquium for the Philosophy of Science 1969–1972*. 1974.
15. Robert S. Cohen, J. J. Stachel, and Marx W. Wartofsky (eds.), *For Dirk Struik. Scientific, Historical and Political Essays in Honor of Dirk Struik*. 1974.
16. Norman Geschwind, *Selected Papers on Language and the Brain*. 1974
17. B. G. Kuznetsov, *Reason and Being: Studies in Classical Rationalism and Non-Classical Science*. 1987
18. Peter Mittelstaedt, *Philosophical Problems of Modern Physics*. 1976
19. Henry Mehlberg, *Time, Causality, and the Quantum Theory* (2 vols.). 1980.

20. Kenneth F. Schaffner and Robert S. Cohen (eds.), *Proceedings of the 1972 Biennial Meeting, Philosophy of Science Association.* 1974
21. R. S. Cohen and J. J. Stachel (eds.), *Selected Papers of Léon Rosenfeld.* 1978.
22. Milic Čapek (ed.), *The Concepts of Space and Time. Their Structure and Their Development.* 1976.
23. Marjorie Grène, *The Understanding of Nature, Essays in the Philosophy of Biology.* 1974.
24. Don Ihde, *Technics and Praxis. A Philosophy of Technology.* 1978.
25. Jaakko Hintikka and Unto Remes, *The Method of Analysis. Its Geometrical Origin and Its General Significance.* 1974.
26. John Emery Murdoch and Edith Dudley Sylla, *The Cultural Context of Medieval Learning.* 1975.
27. Marjorie Grene and Everett Mendelsohn (eds.), *Topics in the Philosophy of Biology.* 1976.
28. Joseph Agassi, *Science in Flux.* 1975.
29. Jerzy J. Wiatr (ed.), *Polish Essays in the Methodology of the Social Sciences.* 1979.
30. Peter Janich, *Protophysics of Time.* 1985.
31. Robert S. Cohen and Marx W. Wartofsky (eds.), *Language, Logic and Method.* 1983.
32. R. S. Cohen, C. A. Hooker, A. C. Michalos, and J. W. van Evra (eds.), *PSA 1974: Proceedings of the 1974 Biennial Meeting of the Philosophy of Science Association.* 1976.
33. Gerald Holton and William Blanpied (eds.), *Science and Its Public: The Changing Relationship.* 1976.
34. Mirko D. Grmek (ed.), *On Scientific Discovery.* 1980.
35. Stefan Amsterdamski, *Between Experience and Metaphysics. Philosophical Problems of the Evolution of Science.* 1975.
36. Mihailo Marković and Gajo Petrović (eds.), *Praxis, Yugoslav Essays in the Philosophy and Methodology of the Social Sciences.* 1979.
37. Hermann von Helmholtz, *Epistemological Writings. The Paul Hertz/Moritz Schlick Centenary Edition of 1921 with Notes and Commentary by the Editors.* (Newly translated by Malcolm F. Lowe. Edited, with an Introduction and Bibliography, by Robert S. Cohen and Yehuda Elkana). 1977.
38. R. M. Martin, *Pragmatics, Truth, and Language.* 1979.
39. R. S. Cohen, P. K. Feyerabend, and M. W. Wartofsky (eds.), *Essays in Memory of Imre Lakatos.* 1976.
40. B. M. Kedrov and V. Sadovsky. *Current Soviet Studies in the Philosophy of Science.* Forthcoming.
41. M. Raphael, *Theorie des Geistigen Schaffens auf Marxistischer Grundlage.* Forthcoming.
42. Humberto R. Maturana and Francisco J. Varela, *Autopoiesis and Cognition. The Realization of the Living.* 1980.
43. A. Kasher (ed.), *Language in Focus: Foundations, Methods and Systems. Essays Dedicated to Yehoshua Bar-Hillel.* 1976.
44. Trân Duc Thao, *Investigations into the Origin of Language and Consciousness.* (Translated by Daniel J. Herman and Robert L. Armstrong; edited by Carolyn

R. Fawcett and Robert S. Cohen). 1984.
45. A. Ishimoto (ed.), *Japanese Studies in the History and Philosophy of Science.*
46. Peter L. Kapitza, *Experiment, Theory, Practice.* 1980.
47. Maria L. Dalla Chiara (ed.), *Italian Studies in the Philosophy of Science.* 1980.
48. Marx W. Wartofsky, *Models: Representation and the Scientific Understanding.* 1979.
49. Trân Duc Thao, *Phenomenology and Dialectical Materialism.* 1985.
50. Yehuda Fried and Joseph Agassi, *Paranoia: A Study in Diagnosis.* 1976.
51. Kurt H. Wolff, *Surrender and Catch: Experience and Inquiry Today.* 1976.
52. Karel Kosik, *Dialectics of the Concrete.* 1976.
53. Nelson Goodman, *The Structure of Appearance.* (Third edition). 1977.
54. Herbert A. Simon, *Models of Discovery and Other Topics in the Methods of Science.* 1977.
55. Morris Lazerowitz, *The Language of Philosophy. Freud and Wittgenstein.* 1977.
56. Thomas Nickles (ed.), *Scientific Discovery, Logic, and Rationality.* 1980.
57. Joseph Margolis, *Persons and Minds. The Prospects of Nonreductive Materialism.* 1977.
58. G. Radnitzky and G. Andersson (eds.), *Progress and Rationality in Science,* 1978.
59. Gerard Radnitzky and Gunnar Andersson (eds.), *The Structure and Development of Science.* 1979.
60. Thomas Nickles (ed.), *Scientific Discovery: Case Studies.* 1980.
61. Maurice A. Finocchiaro, *Galileo and the Art of Reasoning.* 1980.
62. William A. Wallace, *Prelude to Galileo.* 1981.
63. Friedrich Rapp, *Analytical Philosophy of Technology.* 1981.
64. Robert S. Cohen and Marx W. Wartofsky (eds.), *Hegel and the Sciences.* 1984.
65. Joseph Agassi, *Science and Society.* 1981.
66. Ladislav Tondl, *Problems of Semantics.* 1981.
67. Joseph Agassi and Robert S. Cohen (eds.), *Scientific Philosophy Today.* 1982.
68. Władysław Krajewski (ed.), *Polish Essays in the Philosophy of the Natural Sciences.* 1982.
69. James H. Fetzer, *Scientific Knowledge.* 1981.
70. Stephen Grossberg, *Studies of Mind and Brain.* 1982.
71. Robert S. Cohen and Marx W. Wartofsky (eds.), *Epistemology, Methodology, and the Social Sciences.* 1983.
72. Karel Berka, *Measurement.* 1983.
73. G. L. Pandit, *The Structure and Growth of Scientific Knowledge.* 1983.
74. A. A. Zinov'ev, *Logical Physics.* 1983.
75. Gilles-Gaston Granger, *Formal Thought and the Sciences of Man.* 1983.
76. R. S. Cohen and L. Laudan (eds.), *Physics, Philosophy and Psychoanalysis.* 1983.
77. G. Böhme et al., *Finalization in Science,* ed. by W. Schäfer. 1983.
78. D. Shapere, *Reason and the Search for Knowledge.* 1983.
79. G. Andersson, *Rationality in Science and Politics.* 1984.
80. P. T. Durbin and F. Rapp, *Philosophy and Technology.* 1984.
81. M. Marković, *Dialectical Theory of Meaning.* 1984.

82. R. S. Cohen and M. W. Wartofsky, *Physical Sciences and History of Physics.* 1984.
83. E. Meyerson, *The Relativistic Deduction.* 1985.
84. R. S. Cohen and M. W. Wartofsky, *Methodology, Metaphysics and the History of Sciences.* 1984.
85. György Tamás, *The Logic of Categories.* 1985.
86. Sergio L. de C. Fernandes, *Foundations of Objective Knowledge.* 1985.
87. Robert S. Cohen and Thomas Schnelle (eds.), *Cognition and Fact.* 1985.
88. Gideon Freudenthal, *Atom and Individual in the Age of Newton.* 1985.
89. A. Donagan, A. N. Perovich, Jr., and M. V. Wedin (eds.), *Human Nature and Natural Knowledge.* 1985.
90. C. Mitcham and A. Huning (eds.), *Philosophy and Technology II.* 1986.
91. M. Grene and D. Nails (eds.), *Spinoza and the Sciences.* 1986.
92. S. P. Turner, *The Search for a Methodology of Social Science.* 1986.
93. I. C. Jarvie, *Thinking about Society: Theory and Practice.* 1986.
94. Edna Ullmann-Margalit (ed.), *The Kaleidoscope of Science.* 1986.
95. Edna Ullmann-Margalit (ed.), *The Prism of Science.* 1986.
96. G. Markus, *Language and Production.* 1986.
97. F. Amrine, F. J. Zucker, and H. Wheeler (eds.), *Goethe and the Sciences: A Reappraisal.* 1987.
98. Joseph C. Pitt and Marcella Pera (eds.), *Rational Changes in Science.* 1987.
99. O. Costa de Beauregard, *Time, the Physical Magnitude.* 1987.
100. Abner Shimony and Debra Nails (eds.), *Naturalistic Epistemology: A Symposium of Two Decades.* 1987.
101. Nathan Rotenstreich, *Time and Meaning in History.* 1987.
102. David B. Zilberman (ed.), *The Birth of Meaning in Hindu Thought.* 1987.
103. Thomas F. Glick (ed.), *The Comparative Reception of Relativity.* 1987.
104. Zellig Harris *et al., The Form of Information in Science.* 1987
105. Frederick Burwick, *Approaches to Organic Form: Permutations in Science and Culture.* 1987.
106. M. Almási, *Philosophy of Appearances.* Forthcoming.
107. S. Hook, W. L. O'Neill, and R. O'Toole, *Philosophy, History and Social Action. Essays in Honor of Lewis Feuer.* 1988.
108. I. Hronszky, M. Fehér, and B. Dajka (eds.), *Scientific Knowledge Socialized. Selected Proceedings of the Fifth Joint International Conference on History and Philosophy of Science Organized by the IUHPS, Veszprém, 1984.* Forthcoming.
109. P. Tillers and E. D. Green (eds.), *Probability and Inference in the Law of Evidence. The Uses and Limits of Bayesianism.* 1988.
110. E. Ullmann-Margalit (ed.), *Science in Reflection. The Israel Colloquium: Studies in History, Philosophy, and Sociology of Science.* 1988.
111. K. Gavroglu, Y. Goudaroulis, and P. Nicolacopoulos (eds.), *Imre Lakatos and Theories of Scientific Change.* 1989.
112. Barry Glassner and Jonathan D. Moreno (eds.), *The Qualitative-Quantitative Distinction in the Social Sciences.* 1989.
113. K. Arens, *Structures of Knowing: Psychologies of the Nineteenth Century.* 1989.

114. A. Janik, *Style, Politics and the Future of Philosophy*. 1989.
115. F. Amrine (ed.), *Literature and Science as Modes of Expression*. 1989.
116. James Robert Brown and Jürgen Mittelstrass (eds.), *An Intimate Relation: Studies in the History and Philosophy of Science Presented to Robert E. Butts on his 60th Birthday*. 1989.
117. F. D'Agostino and I. C. Jarvie (eds.), *Freedom and Rationality: Essays in Honor of John Watkins*. 1989.
118. D. Zolo, *Reflective Epistemology: The Philosophical Legacy of Otto Neurath*. 1989.
119. Michael Kearn, Bernard S. Phillips and Robert S. Cohen (eds.), *George Simmel and Contemporary Sociology*. 1989.
120. Trevor H. Levere and William R. Shea (eds.), *Nature, Experiment, and the Sciences: Essays on Galileo and the History of Science in Honour of Stillman Drake*. 1989.
121. P. Nicolacopoulos (ed.), *Greek Studies in the Philosophy and History of Science*. 1990.
122. R. Cooke and D. Constantini (eds.), *Statistics in Science. The Foundations of Statistical Methods in Biology, Physics and Economics*. 1990